Streaming

Streaming

Movies, Media, and Instant Access

Wheeler Winston Dixon

The University Press of Kentucky

Scholarly publisher for the Commonwealth,
serving Bellarmine University, Berea College, Centre
College of Kentucky, Eastern Kentucky University,
The Filson Historical Society, Georgetown College,
Kentucky Historical Society, Kentucky State University,
Morehead State University, Murray State University,
Northern Kentucky University, Transylvania University,
University of Kentucky, University of Louisville,
and Western Kentucky University.
All rights reserved.

Editorial and Sales Offices: The University Press of Kentucky
663 South Limestone Street, Lexington, Kentucky 40508-4008
www.kentuckypress.com

17 16 15 14 13   5 4 3 2 1

Library of Congress Cataloging-in-Publication Data

Dixon, Wheeler W., 1950–
  Streaming : movies, media, and instant access / Wheeler Winston Dixon.
    pages   cm
  Includes bibliographical references and index.
  ISBN 978-0-8131-4217-3 (hardcover : alk. paper)—
  ISBN 978-0-8131-4218-0 (epub) — ISBN 978-0-8131-4219-7
  (pbk. : alk. paper) — ISBN 978-0-8131-4224-1 (pdf)
  1. Streaming technology (Telecommunications) 2. Digital video. 3. Business enterprises—Computer networks. I. Title.
  TK5105.386.D59 2013
  006.7'876—dc23

                    2012051381

This book is printed on acid-free paper meeting
the requirements of the American National Standard
for Permanence in Paper for Printed Library Materials.

Manufactured in the United States of America.

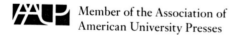 Member of the Association of
American University Presses

For Gwendolyn,
My One and Only

We are on a shift that is as momentous and as fundamental as the shift to the electrical grid. It's happening a lot faster than any of us thought.

—*Andrew R. Jassy, as quoted by Quentin Hardy*

Contents

1

On Demand

There can be no doubt that the digitization of the moving image has radically and irrevocably altered the phenomenon we call the cinema, and that the characteristics of this transformation leave open an entirely new field of visual figuration. For those who live and work in the postfilmic era—that is, those who have come to consciousness in the last twenty years—the digital world is not only an accomplished fact but also the dominant medium of visual discourse. Many observers have remarked that the liberation of the moving image from the tyranny of the "imperfect" medium of film is a technical shift that is not only inevitable but also desirable. And this tectonic shift in cinema is only part of the overall digitization of society.

This shift to digital cinema is now under way around the world, and much of it still resides in Hollywood, which retains an almost hegemonic grip on international image discourse. The way the image is captured, disseminated, and consumed by contemporary audiences, and the manner in which this process, or series of processes, is constantly being revised, means that the medium is at a genuine "tipping point" in its history. The same holds true for music, texts, anything that can be streamed. This switch to digital will bring about a permanent change in the habits of viewers, who can now see films on everything from cell phones to conventional theater screens.

It is an inescapable fact that we will soon experience a complete changeover to digital formatting, eschewing film entirely, and many audience members are deeply disturbed by the thought, as if, in losing the platform of film, we will be losing some essential essence of the medium. But a moving image is simply that, as Godard long ago demonstrated with his revolutionary *mixages* of film and video in *Histoires du Cinéma*, and what was once conjecture has now become an accomplished fact.

Film has vanished, but the image remains, albeit in a new, sleeker format. It isn't a question of whether this is good or bad; it's simply a fact. The movies have changed, and we are changing with them. *Streaming: Movies, Media, and Instant Access* documents part of that process. The shift to a digital cinema will not be without controversy. Film comes with one set of values inherently present in the stock itself (a tendency toward warmth in color for some film stocks, or toward cooler hues in others, as well as characteristics of grain, depth, and definition that are unique to each individual film matrix), while the digital video image offers another entirely different set of characteristics, verging on a hyperreal glossiness that seems to shimmer on the screen. To achieve a reconsideration of the basic states of representationalism inherent in any comparison of these two mediums is a difficult task, calling into question more than a century of cinematic practice and a host of assumptions shared by practitioners and viewers alike. But it is work that needs to be done, and this text represents a step in that direction.

I have been gathering the material for this book for several years, and during that time, it has become clear that streaming—for movies, music, books, and, indeed, every form of textual material—has become the dominant mode of delivery. It's also becoming faster than ever before. The most recent breakthrough, in late June 2012, was "twisted light" technology, which can, as noted in the scientific web journal *The Bunsen Burner*, "transmit upwards of 2.56 terabits of data per second using twisted beams of light. . . . It may one day be commonplace to download data packages the equivalent of 70 DVDs in one second. . . . The test resulted in data transfer speeds 85,000 times faster than broadband internet speeds, some of the fastest ever recorded." DVDs, on-demand cable movies, CDs, and other platforms continue to hang on, but as niche products. For today's audiences, everything has to be instant—it's now or never. Netflix built its business model on delivering DVDs to the doorsteps of its customers; this effectively killed Blockbuster Video's brick-and-mortar "go to the store" model, just as Amazon did with books, thus wiping out a large number of independent bookstores, even large chains, throughout the world.

The concept of "streaming," of course, is nothing new. Early television was essentially streaming video, raw and unprocessed, with plenty of airtime and minimal commercial interruption. I'm thinking particularly of *Open End*, a television talk show that aired from 1958 to 1987—quite a respectable run for any show. Its initial incarnation, from 1958 to 1961 on channel 13, WNTA New York (before the station became a PBS affiliate and, indeed, before PBS existed), was unique because the show really *was* open-ended. Host David Susskind would gather philosophers, performers, authors, theologians, politicians—really, anyone who was interested in intelligent discourse—and let them chat about whatever came into their heads until everyone ran out of steam. *Open End* sometimes ran for two hours or so; sometimes it would go on until 3:00 in the morning, and on one memorable occasion, it didn't end until nearly 5:00 AM. Susskind would pick a topic for the evening, but the discussion would soon range into whatever the participants wished to talk about, at whatever length they wished to speak. No one cut anyone off, only one person spoke at a time, and voices were raised only on rare occasions. Susskind was conscientious in ensuring that all points of view—political and philosophical—were represented by his numerous guests. There were always five or six participants per show, and it made for lively and informative television. And, of course, there were no commercials.

Can you imagine such a program on television today? "We're almost out of time," "We have only a few moments left," "Quickly before we go," "In our remaining moments"—in our *remaining moments*? Are we about to *die* or something? We have plenty of time to think, to talk, to discuss, to examine, to air alternative points of view, but we just don't do it anymore. Talk shows have degenerated into shouting matches, and guests participate only if they have a movie, book, or website to plug. Real intellectual exchanges have been forgotten. When channel 13 joined the PBS network, *Open End* moved to channel 5, WNEW TV, and was chopped to a two-hour format with commercials; it was renamed *The David Susskind Show* and ran from 1961 until Susskind's death in 1987. The new format represented a distinct loss, but still, Susskind kept the guest lists interesting and the topics timely, and he maintained an open space for relaxed discussion. It's our loss that we have nothing like this on television today, but in some respects, *Open End* not only anticipated streaming live video channels; it also demonstrated how fascinating and informative they could be.

Today, the choice between streaming and hard-copy technology is the major divide in consumer viewing habits. On November 22, 2010, Netflix announced it would offer a streaming-only service to viewers and simultaneously hike the subscriber price for DVDs rented through the mail; although the company quickly abandoned the "separate service" plan, it's clear that Netflix wants to do away with DVDs altogether. As Netflix's CEO, Reed Hastings, said in a statement of company policy on the date of the announcement, "We are now primarily a streaming video company delivering a wide selection of TV shows and films over the Internet" (as quoted in Edwards and Rabil). At the time, Wall Streeters expected this two-tier deal to succeed, providing yet another example of how out of touch business professionals often are with consumer preferences. *Bloomberg News* reported:

> The price increase lets Netflix pass on rising postage costs and protect average revenue per customer as more subscribers choose streaming only, which is more profitable. . . . The streaming-only plan, priced at $7.99 a month, had been tested in the U.S. after a similar option started in Canada two months ago surpassed expectations. . . . The subscription price for unlimited streaming and one mail-order DVD at a time will increase to $9.99 a month from $8.99, Netflix said. The cost to have more DVDs at a time will also go up. The price change takes effect now for new customers and in January [2012] for existing customers. (Edwards and Rabil)

But of course, the backlash was immediate and pronounced, and the plan was soon abandoned.

There's another major question lurking here: what about all the classic films that aren't available as streaming video? In essence, they will cease to exist. Netflix is banking on the fact that most people have no real knowledge of film history, so they'll content themselves with streaming only the most recent and popular films. This is something akin to Amazon deciding to do away with physical books altogether and offering everything only on Kindle. No doubt Amazon is thinking about this possibility and would love to do it, but this would marginalize hundreds of thousands of books. Netflix is doing the same thing with films.

As one immediate example, none of Frederick Wiseman's documentaries are available through Netflix, despite his repeated attempts to get Netflix to distribute them. It's not the financial terms holding Netflix back; the company simply doesn't want to be bothered with a small-timer like Wiseman, who self-distributes his work through his own company, Zipporah Films. You can buy DVDs of some of the more recent Wiseman films on Amazon, but for his earlier films, Amazon's website offers just *books* that *describe* the films—not the films themselves.

Thus, after killing off all the brick-and-mortar stores for DVDs, CDs, and books, Amazon and Netflix seem poised to do away with all vestiges of the real and enter the digital-only domain. What will be lost in the process is not only the physical reality of books and DVDs; many titles won't make it to Kindle or streaming video, simply because they're not popular enough. In short, we'll have the "top ten" classics, and the rest of film history—many superb, remarkable films—will gather dust on the shelf. If you can just click and stream, why wait for the mailman?

In addition, there is a plethora of programming designed specifically for the web. Some programs come and go like mayflies and die a quick death; others build up a long-term audience and keep coming back year after year to a cadre of loyal viewers. *Web Therapy*, for example, has now amassed forty-six episodes, and Meryl Streep recently guest-starred in a three-episode story arc. Syfy Television (formerly Sci-fi, until the need to copyright the channel's name forced the somewhat awkward switch to Syfy) has been churning out ten-minute segments of a web serial entitled *Riese*, with an eye to combining them into a two-hour TV pilot for the network. Oddly, Showtime has created an animated web companion for its hit live-action show *Dexter*, about a serial killer. Entitled *Dark Echo*, the web program offers brief segments (three to six minutes) of additional backstory on *Dexter* for its numerous devotees (Hale).

With the switch to streaming video, a whole host of royalty, content, and distribution issues needs to be addressed. Even television shows cost money to stream to consumers; in fact, they often cost much more than theatrical feature films. As Richard Kastelein observes:

> In 2010, streamed videos outnumbered DVD rentals for the first time in Netflix's history. . . . But the online move has cost Netflix at least $1.2 billion, according to *CNET*. That's the amount Netflix has committed to paying Hollywood studios for the rights to stream their movies and TV shows. And it's up from $229 million three months ago, the company disclosed in an SEC filing [on October 28, 2010]. Most of that leap comes from a five-year deal that Netflix previously announced with the Epix pay channel, which is thought to be in the $900 million to $1 billion range. But that number could jump again within the next year, when Netflix's deal with the Starz pay channel expires.

Nor is this cost likely to decrease in the future. Analysts from Hudson Square Research Group note that, "given that major studios are very likely to expect increasing rates to obtain streaming only rights, we fully expect Netflix's content costs to rise further" (as quoted in Hayden). Moreover, despite the fact that "Netflix has been aggressively acquiring streaming content [and in 2011 alone] signed licensing deals with NBC Universal, Warner Bros., 20th Century Fox, Epix, Relativity Media and Nu Image/Millennium Films" (Seitz), the price for these deals is bound to increase in the future, and this will have to be passed on to consumers.

But Netflix is hardly alone in its quest for digital dominance. As Patrick Seitz notes, "Netflix isn't the only game in town. Besides Redbox, Hulu just launched a premium streaming video service for $7.99 a month. Deep-pocketed companies like Apple,

Amazon, Google and Wal-Mart are aiming to take a slice of the market as well." Recently, YouTube committed $100 million to new programming created exclusively for the channel, supplanting the old slogan "Broadcast Yourself." New channels on the site, such as *Young Hollywood*, which deals in celebrity gossip, will replace the vox populi with pop stars.

Music, books, movies: they've all gone streaming, and it's well past the tipping point of being a fringe market. It's *the* market; it's easy to access and requires no storage space—you can store your books, movies, and music in clouds. That's all there is to it. This also demonstrates, to me at least, that streaming is now the dominant technology, and corporations are taking notice; it's where the money is. A side effect of all this, of course, is that DVD players—even the relatively new, highly touted Blu-ray players—will become obsolete in an all-streaming world. The industry continues to roll out the hardware, but will consumer demand be there, even with the added "lure" of 3-D?

But as events have proved, with streaming technology, even when you buy something, you don't *really* own it. An early harbinger of this was Amazon's now famous (and all too obviously ironic) deletion of George Orwell's *1984* from Kindles around the United States when the company discovered it had accidentally uploaded an unauthorized version of the novel. As Brad Stone wrote on July 17, 2009 (light years ago, by today's standards):

> In a move that angered customers and generated waves of online pique, Amazon remotely deleted some digital editions of the books from the Kindle devices of readers who had bought them. An Amazon spokesman, Drew Herdener, said in an e-mail message that the books were added to the Kindle store by a company that did not have rights to them, using a self-service function. "When we were notified of

this by the rights holder, we removed the illegal copies from our systems and from customers' devices, and refunded customers' [money]. We are changing our systems so that in the future we will not remove books from customers' devices in these circumstances," Mr. Herdener said.

Although Amazon has promised to make changes, the very fact that it possesses the technology to simply remove books or anything else from your Kindle, or perhaps even your computer, at will—truer now than ever, if you use cloud storage—is unsettling. As David Bowie observed in a conversation with Jon Pareles in 2002, long before the music industry as a whole figured it out and started demanding "360 deals" from performers, covering all aspects of their merchandisability:

> The absolute transformation of everything that we ever thought about music will take place within 10 years, and nothing is going to be able to stop it. I see absolutely no point in pretending that it's not going to happen. . . . Music itself is going to become like running water or electricity. . . . You'd better be prepared for doing a lot of touring because that's really the only unique situation that's going to be left. It's terribly exciting. But on the other hand it doesn't matter if you think it's exciting or not; it's what's going to happen.

There's also the issue of platforms: how many different sources of images are there, and who will win the war for commercial domination? Recently, the major studios have embraced UltraViolet, a system that will seemingly revolutionize (yet again) content delivery systems, and they have teamed up with Wal-Mart to deliver films to viewers. According to Michelle Kung and Miguel Bustillo in the *Wall Street Journal*:

The UltraViolet system, which has been slow getting off the ground, is a digital "proof of purchase" system that allows a consumer to store movie or TV titles in a free, online personal library. Once a video has been added to the Ultra-Violet Library it can be streamed over the web or downloaded for viewing on a computer, TV, or a range of mobile devices. UltraViolet, which is backed by a group of more than 70 entertainment, technology and retail companies, was announced in January 2011, but so far has only about one million users in the country.

. . . That's where Wal-Mart comes in. Employees of Wal-Mart will help customers create UltraViolet accounts, according to the people familiar with the plan. Wal-Mart staff will also check DVDs that shoppers already own, adding titles that are part of [the] UltraViolet system to their accounts for a small fee.

With a rollout like this, the system has a real chance to make a dent in the market. Television ads for the service are already airing on a national basis. Yet as industry analyst Jim Taylor points out, with UltraViolet, only the authorization code for a film is stored "in the cloud and accessed using the Internet . . . the UltraViolet system itself doesn't store or deliver content. Retailers, download service providers (DSPs), and streaming service providers (LASPs) store and deliver your movies." So theoretically, the system is endlessly expandable, but it seems clear to me that only the most commercial films and television programs will make the cut.

And then there's this to consider: when a film is digitized, it requires constant maintenance, and transfer to new platforms, to ensure its continued existence, as demonstrated in a prescient 2007 book-length report entitled *The Digital Dilemma: Strate-*

gic Issues in Archiving and Accessing Digital Motion Picture Materials, commissioned and executed by the Science and Technology Council of the Academy of Motion Picture Arts and Sciences. Milton Shefter, a pioneering film archivist and author of the report, convincingly argued that despite its pristine visual quality, the digital image is inherently unstable and in a constant state of flux, and storage of digital imagery is a never-ending process of maintenance and upgrading. As Michael Cieply noted in a 2007 article in the *New York Times*:

> To store a digital master record of a movie costs about $12,514 a year, versus the $1,059 it costs to keep a conventional film master. Much worse, to keep the enormous swarm of data produced when a picture is "born digital"— that is, produced using all-electronic processes, rather than relying wholly or partially on film—pushes the cost of preservation to $208,569 a year [that's right—$208,569 *per year,* which is] vastly higher than the $486 it costs to toss the equivalent camera negatives, audio recordings, on-set photographs and annotated scripts of an all-film production into the cold-storage vault.

Indeed, the major studios realize that digital elements are inherently unstable and that pulling them from storage for re-release is often fraught with difficulty. That's why they routinely keep a 35mm fine-grain negative and positive print of each new release, whether it was produced on film or not, in their vaults as a backup in case something goes wrong with the digital master. And now, more than ever, studios' back catalogs are of increasing importance. It's easy to see that the rush to streaming creates many new ways to "monetize" existing films, music, and books in a way that perpetuates the consumer's dependence on the supplier and ensures a constant revenue stream not only to access but also to store, download, or view the works one has supposedly "purchased." And as streaming downloads of one variety or another become more and more common, it seems that the future of the DVD is uncertain at best. As Matthew Calamia reported on June 1, 2011:

> Apple is in talks with movie companies, including Warner Bros. and 20th Century Fox, according to *CNET*, to allow Apple to stream movies its customers own through Apple's iCloud streaming service, which will be unveiled next week at the company's Worldwide Developers Conference. Streaming movies would work the same way as streaming music. With a user's permission, the iCloud would search through an iTunes library for previously owned movies, and then enable those movies to stream wirelessly. For example, a customer who owns *The King's Speech* on their desktop or laptop computer would get the store's permission before being able to stream that movie from the cloud on any device. . . . Social network giant Facebook has also begun renting movies on its site as well, rounding out an early list of players in [what] will likely become a booming market.

So will Netflix be around to service its customers in the future? If not, what will take its place? It doesn't really matter, in the long run; the whole battle over marketplace dominance is simply what the trade journal *Variety* used to call the "executive shuffle." Although the "platform" of the DVD may vanish, the "films" themselves will remain. Audiences have adjusted to viewing moving images in a variety of different ways. They may still want to see their dreams and desires projected on the large screen for the visceral thrill of the spectacle, as well as the communal aspect of any public performance. But who knows what form the theatrical film experience will take in the future?

Indeed, many twentieth-century, and even twenty-first-century, technologies we currently take for granted will soon disappear. One example is regular mail—soon all our correspondence will be connected electronically, and only the shipping of physical objects will survive. Home telephones will be replaced by "mini computers the size of all phones," as landlines are discontinued. "CDs, DVDs, Blu-rays, thumb drives, video game discs . . . and their players" will become obsolete as music, movies, games, and text are streamed directly from the web. Other items on the way out include business cards, conventional road maps, cash, analog clocks, phone books, bank deposit slips, and incandescent lightbulbs. Even the passwords to your various electronic devices will become unnecessary "as facial biometric software will enable you to simply look at your cell phone or tablet—and unlock it" (Towner).

With more films, videos, television programs, and Internet films being produced than ever before, and with international image boundaries crumbling thanks to the pervasive influence of the World Wide Web, the coming years will bring an explosion of voices from around the globe. In a more democratic process, even the most marginalized factions of society will have a voice, no matter what new business models are brought into play. As Wikipedia's "blackout" of January 18, 2012, proved, the web has a great ability to democratize itself and to make its intentions known to the public, to businesspeople, and to legislators.

Why was the blackout necessary? Because the normal channels of communication and persuasion weren't working. On November 15, 2011, numerous websites and social networking platforms—AOL, Google, Reddit, Wikipedia, Mozilla, Word Press, MoveOn.org, and others—sent a joint letter to Patrick

Leahy, chairman of the Senate Judiciary Committee, as well as to congressmen Chuck Grassley, Lamar Smith, and John Conyers Jr., expressing their "concern with legislative measures that have been introduced in the United States Senate and United States House of Representatives, S. 968 (the 'Protect IP Act') and H.R. 3261 (the 'Stop Online Piracy Act')." The signers of the letter state that while they "support the bills' stated goals—providing additional enforcement tools to combat foreign 'rogue' websites that are dedicated to copyright infringement or counterfeiting"— they point out that "the bills as drafted would expose law-abiding U.S. Internet and technology companies to new uncertain liabilities, private rights of action, and technology mandates that would require monitoring websites" (AOL et al.). In the industry's view, as Declan McCullagh notes, the whole problem boils down to "two words: rogue sites." He explains:

> That's Hollywood's term for websites that happen to be located in a nation more hospitable to copyright infringement than the United States is (in fact, the U.S. is probably the least hospitable jurisdiction in the world for such an endeavor). Because the target is offshore, a lawsuit against the owners in a U.S. court would be futile. The U.S. Chamber of Commerce, in a letter to the editor of *The New York Times*, put it this way: "Rogue Web sites . . . steal America's innovative and creative products[,] attract more than 53 billion visits a year[,] and threaten more than 19 million American jobs."

What do the bills propose to do? In the view of their sponsors, the bills are an attempt to stop pirate websites operating outside the United States from providing illegally obtained copyrighted content—movies, music, text, images—through American portals. Naturally, industry spokespersons such as former U.S. senator Christopher Dodd, who now has Jack Valenti's old job as head of the Motion Picture Association of America (MPAA), support both bills vigorously. In a statement on official MPAA letterhead, Dodd blasted the proposed blackout action, vociferously declaring:

> Technology business interests are resorting to stunts that punish their users or turn them into their corporate pawns, rather than coming to the table to find solutions to a problem that all now seem to agree is very real and damaging. It is an irresponsible response and a disservice to people who rely on them for information and use their services. It is also an abuse of power given the freedoms these companies enjoy in the marketplace today. It's a dangerous and troubling development when the platforms that serve as gateways to information intentionally skew the facts to incite their users in order to further their corporate interests. A so-called "blackout" is yet another gimmick, albeit a dangerous one, designed to punish elected and administration officials who are working diligently to protect American jobs from foreign criminals.

Movie piracy is a very real concern that costs the industry hundreds of millions of dollars every year, at the very least. But what's the other side of the coin? While Stephanie Condon states that "Internet companies and their investors would readily say that they're holding the 'blackout' to protect their corporate interests—and the entire burgeoning Internet-based economy," the reality is far more complex. Although such companies as Mozilla, Open DNS, PayPal, Twitter, the Wikipedia Foundation, Yahoo!, Zynga, the Game Network, AOL, eBay, Etsy, Facebook, Four-

Square, Google, LinkedIn, and others opposed the legislation, so did the American Civil Liberties Union, the American Association of Law Libraries, the American Library Association, the American Society of News Editors, the Association of College and Research Libraries, the Brookings Institution, the Center for Democracy, and numerous other groups that aren't really part of the corporate mainstream.

When Wikipedia and the other sites returned online, they all did so with a message to their users, congratulating them on a successful boycott that resulted in the scuttling of the bills—for the moment. Wikipedia's message read in part: "More than 150 million people saw our message asking if you could imagine a world without free knowledge. You said no. You shut down Congress' switchboards. You melted their servers. Your voice was loud and strong. Millions of people have spoken in defense of a free and open internet."

But the struggle continues; it is a battle for profit, on the one hand, and for art and public discourse, on the other. The end result, of course, is unknowable, but for now, streaming and downloading rule, and this technology seems poised to become dominant in the coming months and perhaps years. We're moving away from a hard-media society. In the future, there will be only images stored on hard drives, in clouds, and in our memories, as we watch the digital transformation of the cinema.

And, of course, film itself as a technology is vanishing. A few months ago I was watching Kathryn Bigelow's excellent film *The Hurt Locker* (2008) on DVD, and I was suddenly struck by the fact that it may be one of the last movies actually *shot on film*—in this case, Super 16mm film, with four handheld crews working at once and piling up roughly 200 hours of footage that would eventually be edited down into a 130-minute film. With its rough, raw look, its smash zooms, and its hectic intercutting, mirroring battlefield news photography from the Vietnam War, *The Hurt*

Locker has a visceral reality, especially in its nighttime sequences, that seems to be intrinsically tied to the filmic process. You could have the same images in video, of course, but I don't think the same level of textures and contrasts would be available; you'd get a perfect, pristine, scratch-free image, but a certain richness would be missing. Digital technology simply doesn't have the same spectrum of tonal possibilities, and even though it can mimic millions of different shades of color, the end result is cold, artificial, distant. There's something unreal about it.

When making a film, so to speak, it would be nice to have a choice between using film and going digital. But it seems that the choice has already been made. Aesthetic issues aside, film is being swept into the dustbin of history. As Richard Verrier reported in the *Los Angeles Times* on October 25, 2011, Birns & Sawyer, the oldest renter of film equipment in Hollywood, has thrown in the towel on film—everything's gone digital. As Verrier wrote:

> Call it film's last gasp. Birns & Sawyer, the oldest movie camera rental shop in Hollywood, made history last week when it auctioned off its entire remaining inventory of 16- and 35-mm film cameras. Owner and cinematographer Bill Meurer said he didn't want to part with the cameras, but had little choice as the entertainment industry has largely gone digital. "People aren't renting out film cameras in sufficient numbers to justify retaining them," Meurer said in an interview at his North Hollywood warehouse, where he rents out cameras, lenses, lighting equipment and grip trucks. "Initially, I felt nostalgic, but 95% of our business is digital. We're responding to the market."

As far back as 2000, while lecturing in Stockholm, Sweden, I predicted this shift would happen—not with much enthusiasm, but simply as a matter of fact. At the time, there was one digital theater in New York, and the executives made a big show of dumping 35mm film canisters into a trash bin as a demonstration of their embrace of digital technology. It made for an apt if distressing image: film was heading for the dump. An audience member replied that what happened in one small theater in New York couldn't possibly threaten the hegemony of film production and exhibition; it was simply too ubiquitous and too ingrained. There were millions of 35mm features. And 35mm was about to become obsolete? Ridiculous. I also appeared on an NPR talk show with director Bennett Miller around the same time; I predicted digital would replace film within five years, and everyone in the room thought I was crazy. It's taken a bit longer than that, but the day of all-digital production is here.

And with this shift, of course, comes *digital projection* and a whole new level of studio control. Once upon a time, when you screened a film at a theater, you took the 35mm print out of the shipping case; threaded it up; checked the aspect ratio, focus, and sound level; and ran the film. If you wanted to do an additional screening for a critic or add an extra show, you could. If you wanted to switch the movie from one screen to another in your theater, you could. In short, you had some measure of control over the projection of the films you screened. Not anymore. You now get a digital cinema package, or DCP, for each film, and with it a series of encryption codes, called KDMs, that must be used to "unlock" the digital files for projection, often within a window as short as four hours. Switching screens or adding additional shows now has to be cleared with the distributor every time, usually by an exchange of e-mails. You can't just pull the film off the shelf and run it. Every showing has to be approved, and the digital files must be unlocked with a KDM, on a case-by-case basis. This excerpt from "Digital Cinema Technology: Frequently Asked Questions" explains the process:

KDM is the acronym for Key Delivery Message. The security key for each movie is delivered in a unique KDM, one KDM per digital cinema server. The security key is encrypted within the KDM, which means that the delivery of a KDM to the wrong server or wrong location will not work, and thus such errors cannot compromise the security of the movie. The KDM is a small file, and is typically emailed to the exhibitor. To create the correct KDM, however, requires knowledge of the digital certificate in the projection system's media block.

KDMs have only a *few* [emphasis added] conditions associated with their use:

• A KDM will only work for one movie title on one server.

• A KDM will only work within the prescribed engagement time period.

• To play a movie on two servers requires two KDMs for the movie. This means that to move a movie to a 2nd server requires a 2nd KDM. The engagement time window of the KDM is set per the business requirements of the studio distributing the movie. If your KDM expires and you don't have a new KDM to continue on the engagement, then you cannot play the movie.

This is about studio control, nothing more. It takes all authority away from the exhibitor; it's a hypersurveillance system that comes from the top down and limits what theater owners can do. Digital projection has many significant attributes—superior picture and sound; no scratches; clean, crisp images—but now movies don't really exist unless they're unlocked by the KDM, and they have no portability. This is what the studios want; it's what theorist Tim Wu describes as the desire to control "the master switch"—to have illimitable power over one's domain. It's not good for the public or critics or exhibitors—a real measure of discretionary freedom has been lost. And there's another factor, as any theatrical projectionist can tell you: DCPs and KDMs are notoriously unreliable.

Recently, there was a very public demonstration of just how unreliable the DCP/KDM system is; it happened at, of all places, the fiftieth New York Film Festival. Director Brian De Palma brought a DCP copy of his new film *Passion* (2012) to screen at Alice Tully Hall, but something was amiss with either the DCP itself or the code to unlock it, and the film was never shown, much to De Palma's obvious displeasure. Festival director Richard Peña (who is retiring after this season) had the unpleasant task of dealing with both the audience and De Palma, neither of whom was satisfied with the outcome of events. As Bob Cashill wrote in the web journal *Pop Dose*:

> The New York Film Festival, which got underway last Friday, celebrates its 50th anniversary this year. I've attended every year since 1994. I've seen dozens of movies there—some extraordinary, many good, some run-of-the-mill, a few terrible. I wondered where Brian De Palma's *Passion*, screened on Saturday, would fit in. I'm still wondering. In what was for me an unprecedented event in my decades of festivalgoing, the screening was cancelled. Why? Three words: Digital Cinema Package, or DCP.
>
> What is DCP? It's heralded as the future of cinema projection, but really it's the present; chances are your local multiplex has gone DCP, as your local independently owned theater or repertory house struggles to find a way to pay for it as celluloid goes up in smoke. The brave new world of digital projection comes with pitfalls, however. Like, if

the system malfunctions, and no one can get a grip on what went wrong, [you're out of luck] as the Film Society of Lincoln Center learned the hard way last night. If you follow the festival at all you've been reading a lot about how Richard Peña, its programming director, is bowing out after 25 years of distinguished service. "I bet he wishes he retired last year," grumbled a fellow patron as we all exited Alice Tully Hall after more than an hour of waiting.

It had been a trying day: apparently a DCP of a Mexican film shown in the afternoon, *Here and There* [*Aquí y allá* (2012), directed by Antonio Mendez Esparza] was plagued by intermittent subtitles that were here and there for half an hour until the problem was fixed. That was nothing compared to this utter fiasco, however. Has a festival presentation ever been cancelled due to a mechanical glitch? Not in the film cans days that I can recall. . . . Peña, who had to keep coming onstage to deliver the worsening news, said that the DCP had been tested without incident minutes before showtime, but minus a code had somehow locked down. Minus someone who could fix the code that was it for the evening. . . . After a half hour or so of waiting Peña announced that audience members who couldn't stay could get refunds at the boxoffice. Only 10–15% seemed to. . . . It was out of Peña's hands, or De Palma's hands, or any human hands. It was a glitch in the machine, a hiccup in the software. And with that the 50th anniversary of the New York Film Festival was tainted.

Nevertheless, digital is taking over, and those 35mm facilities that still exist are facing obsolescence on two fronts: there are fewer and fewer 35mm prints being made available, and since very little 35mm film is being manufactured and used, it's becoming increasingly hard to find parts for 35mm projectors, cameras, or anything else associated with the film medium. Several years ago, I was lucky enough to obtain a 35mm print of Alain Robbe-Grillet's superb, hypnotic first feature *L'immortelle* (1963), directly from the French Cultural Service, and screen it for the students in my class. They were completely enthralled by the experience of seeing the film in its original 35mm format, with all its tonal depth—even though the film is in black and white. But as we watched it, I reminded them that afterward, the film would have to be crated up and sent back to Paris. This was a one-time-only experience. The film is unavailable on DVD, Blu-ray, or even VHS, much less 16mm; it's available only in 35mm. To convert it to a digital master and then release it would apparently cost too much money—even with a down-and-dirty transfer—to make the investment back, so the film will remain in limbo, inaccessible to all but the most dedicated historians. And now, the equipment we use to *project* the film is threatened as well. It is breaking down. Parts are hard to come by, and service technicians are hard to locate. Everyone is headed full force into the future, and no one has time for the past.

Many, many years ago, when I was teaching at Rutgers University, we used to show films—16mm prints—in Scott Hall 123 nearly every night of the week. The screenings were open to all, and the equipment we used was simple in the extreme: a Bell and Howell Model 535 projector with a 1200-watt incandescent lamp that cost about $15 to replace, and some speakers plugged into the projector and placed on the stage in front of the screen. We often ran all-nighters, and we regularly added screenings as audience demand dictated. We watched the films again and again, memorizing them, viewing them—albeit in reduced form—in the medium in which they had been made.

Now, of course, you can stream a lot of films, and the number of available titles increases daily. You can screen them on your laptop, your iPad, and even your fifty-inch plasma TV, but you won't get the experience of seeing the images on film, with all their attendant qualities and defects, and you won't get the communal experience of seeing them with an audience. Movie viewing in the twenty-first century has become, more and more, a solitary vice in which one person tunes out the rest of the world and tunes in to a digitally perfect copy of a film, without having to participate in a group experience.

Does anyone really doubt that the Internet will eventually triumph and smash the rigid program guide that cable and satellite companies shove down our throats? Most of us watch only a few dozen channels regularly, yet we pay for 500. If we could subscribe on a per channel or per show basis, many of us would. It's obvious that the better experience starts with letting people watch what they want, when they want, on whatever device they want. Cable television systems are dinosaurs, trying desperately to hold on in an irrevocably changing media landscape. Streaming is the wave of the future, and now that all the various "on-demand" web services can be run through people's television sets, there's no longer any real barrier to universal audience acceptance. Most viewers today are content to watch on their mobile devices, iPads, or laptops and aren't chained to their flat screens, but for those who are, the web has come home to them. It's all in the family now, and streaming is the future. CBS News, for example, now offers programming exclusively for the web, such as its streaming news show *What's Trending*, with host Shira Lazar; there's already a "Streaming Awards" show for the best in streaming programming, which is streamed, naturally, to viewers around the globe.

Streaming video's price point is also killing the competition. It's a lot *cheaper* to stream a movie than to purchase it, and for most viewers, "watch it and forget it" is a reasonable model of consumption. Collectors may want to hold on to their DVDs for a variety of reasons—some sort of fleeting permanence being one of them—but for most viewers, the idea is to watch it *now* and then move on. With mainstream cinema so bland and predigested, who can disagree? Do you really want to see any of the *Pirates of the Caribbean* films twice or use up valuable storage space in your house or apartment to keep a DVD? When the cost factor between streaming and DVD is so dramatically different, it isn't even a question anymore. As Austin Carr argues, in the case of the *Pirates of the Caribbean* franchise, "*The Curse of the Black Pearl* is going for $17.79 on DVD. On Amazon Instant? Just $1.99. *Dead Man's Chest?* On DVD, $18.99; on Instant, again, $1.99. And *At World's End?* Yes, $1.99 on Instant. On DVD? About $28" ("Streaming Video"). Why would anyone pay more, particularly for something so intrinsically disposable?

In short, we live in a streaming world where everything, inevitably, will be available like running water—all you have to do is pay the utility bill. Of course, there are still large gaps—films, books, and music that haven't entered the streaming era: films that are too marginal or are strictly controlled by their makers; "orphan" films that have no legal copyright status or no living owners and have thus entered the public domain, making them available to anyone but significantly diminishing their potential for corporate profits; out-of-print books or obscure and arcane texts that no one has bothered to scan; music too experimental or resolutely noncommercial for anyone to take notice.

As always, and in all mediums, it's the work created on the margins of society that drives future artistic growth; yet that same work is hard to sell when it's fresh, new, and original. Still, even the most experimental of works may find a place in the streaming universe, which is itself transforming the way we, as viewers, think about the very concept of televisual programming.

As the A. C. Nielsen ratings company noted, surveying viewership in the streaming world is entirely different from determining ratings for programming on conventional television. Nielsen's Jon Gibs told Tim Carmody, "As advertisers, broadcasters and over-the-top providers like Netflix and Hulu are looking at how to address and price new platforms, it's really important to understand how users use and think about each platform. When consumers are looking at content on Netflix, they may be *watching* TV content, but they may not think about it *as* TV content. So how does that change the relationship consumers have to that content—as opposed to Hulu, which is much more clearly seen as a place to watch TV programming, often from just the night before?" How indeed? *Slash Gear* broke down the Nielsen data even further and found that "Nielsen also looked at the people that watch on TV via their computer and found that 14% of Netflix [users] connect this way and 20% of Hulu users. When it comes to the PS3 [PlayStation 3], 15% of Netflix viewers use this method and 3% of Hulu users. Xbox Live streaming has 12% of Netflix and 2% of Hulu. . . . Apple TV has the lowest amount of users with only 1% for each service using it" (McGlaun).

But this situation will certainly change. After all, that's what the streaming landscape is all about: constant permutation. And it's clear that at least some established content providers are latching on to the new technology. CBS's *What's Trending* program is already a web hit, but that's original material. What about CBS's back catalog of shows that were once hits but are no longer on the air? They find a new and vibrant home on the web. In 2011 CBS Entertainment boosted its bottom line by a whopping 97 percent by licensing "select library titles" to Netflix. As Erik Gruenwedel reported on August 2, 2011:

CBS Corp. [on] Aug. 2 said new licensing agreement[s] for the digital streaming of select library titles contributed to its entertainment division increasing second-quarter . . . operating income to $440 million, nearly double the $223 million operating profit during the previous-year period. More importantly, digital revenue from Netflix's subscription-based video-on-demand service does not include recently announced expanded agreements with Netflix for Latin America, the Caribbean and Mexico, and, separately, Amazon Prime. [These deals] "are just another example [of] how we are capitalizing on our content by selling it to new distributors without taking away from established revenue streams," said [CBS] CEO Les Moonves. . . . "We have three *CSI* [programs] that are now on the air and aren't part of these Netflix deals or the Amazon deals," he said. "One day they will come off the air, and we will get a lot of money for them in those platforms."

Once, older television programs went straight to syndication and aired on independent television stations around the world; now, they're streamed.

Even for those films that operate on the margins of economic profitability, streaming video can offer a new haven. Critic David Kehr, for example, notes that "streaming, with its forgivingly low resolution, provides a perfectly acceptable showcase for movies that do not exist in the flawless prints now considered essential for DVD, and particularly Blu-ray, release. There are many movies of interest without reputations or stars big enough to justify the expense of a full-scale digital restoration, but I cling to the conviction that it's better for films to be seen with dust spots or dubious color than not seen at all. Streaming does, or should, open a niche for films that otherwise wouldn't be economically viable." This includes, as Kehr documents, everything

26

from the later films of director Mitchell Liesen to Gene Autry and Roy Rogers westerns, as well as all three seasons of *The Alfred Hitchcock Hour*; films by Japanese masters Kenji Mizoguchi, Mikio Naruse, Kiyoshi Kurosawa, and Seijun Suzuki; relatively obscure titles from such renowned directors as Marcel Carné, Robert Bresson, Mario Monicelli, and Raffaelo Matarazzo; and many other films worthy of any discerning viewer's time and attention. As Kehr notes, at least for those who love the cinema as a medium of artistic expression, "the promise of streaming video as a new platform lies right there: in revalorizing films that don't fit the dominant business model. There's no shortage of movies in this world; what we need are new ways to see them." And streaming video offers us precisely that opportunity, when used to its best advantage.

But at the same time, more and more often, stars are going directly to streaming on the web. As Maria Puente detailed in the summer of 2012:

> The animated sci-fi video series *Electric City*, created and voiced by [Tom] Hanks [can be seen on] Yahoo Screen. [Jerry] Seinfeld's online series, *Comedians in Cars Getting Coffee*, [is streaming through] on Crackle.com. Hanks' 90-minute, 20-episode series, which is airing over three nights, is . . . "the first project in what we call online digital blockbusters," says Erin McPherson, vice president and head of video for Yahoo. "This is new for Yahoo and new for the Internet—this is maiden territory. . . . TV and movies are not going away, but we believe passionately that this is a key part of the future, which is filled with a lot more choices and a lot more mechanisms to watch," she says. "As it becomes more crowded, the unique, exclusive things [like *Electric City*] will stand out."

In Seinfeld's project, Jerry and his colleagues Larry David, Ricky Gervais, Alec Baldwin, and Michael Richards tool around in old cars looking for places to get coffee. (Richards may be hoping for a career comeback on the web, after his disastrous nightclub meltdown was caught on video and streamed around the world, effectively halting his career.) It's a simple enough premise, but then again, Seinfeld is an acknowledged master at creating "something out of nothing," and the series is a substantial hit on Crackle (Puente).

More and more, what's being seen on television on a first-run basis has a "support" show on the web, offering additional content, cheaply created, as an inducement to go online. Thus, people watching the latest iteration of the teleseries *Dallas* on TNT cable television are invited to log in after each episode for an online discussion in which cast members kick around their thoughts on the fictional drama viewers have just witnessed. USA Network's *Covert Affairs* has an adjunct web series entitled *Covert Affairs: Sights Unseen*, which functions as a sort of prequel to the series.

There's also a great deal of brand-new content available. On Yahoo, *Kaboom*, aimed squarely at young men, features former Miss USA Tara Conner in a deliberately provocative show about blowing things up with dynamite—sort of like *Jackass* crossed with cheesecake. There are also videos promoting *William Shatner's Get a Life* documentary on YouTube; Leo Laporte's call-in show *The Tech Guy* on the Twit channel, a sort of *Car Talk* for techies; an original series titled *Burning Love*, documenting the love life of a firefighter; *The Morning After*, which delivers entertainment and celebrity news; *K-Town*, an inside look at Los Angeles's Koreatown neighborhood; and *Spoilers with Kevin Smith*, which deals with contemporary hit movies (Matheson and Maxwell). A lot of it is junk, of course, glorified advertisements at best. But in the midst of all the dross, if you search hard enough, you'll find a good deal of entertainment as well as information that's both enlightening and amusing.

The big hurdle at the moment seems to be the brevity most web viewers expect in their programming; crashing on the sofa and watching a few episodes of Charlie Sheen's new series *Anger Management* while drinking a couple of beers is very different from hunching in front of a laptop or computer screen to watch an episode of the latest web series. They're called "webisodes," of course, and they come in nontraditional lengths—anywhere from ten minutes a pop (as in Seinfeld's new series) to as long as thirty minutes; some are as short as four or five minutes. There's no set standard or format, which is good, but with the commercialization of the web, that will undoubtedly change as dominant forms emerge. Also, many people have hooked up their computers to the family's fifty-two-inch flat screen and are viewing web content in this more relaxed fashion, as well as playing video games, both streaming and conventional. This sort of viewing is rapidly gaining in popularity.

Conceptually, *Anger Management* is an interesting series that, in a way, bridges television and the web. It's shot on a killing schedule of forty-nine pages every two days, with absolutely no rehearsal, to rack up a hundred episodes as quickly as possible for syndication. *Anger Management* aims to achieve that goal in just *two years*, making it a model for hyperaccelerated production as well as economy that should do well on the web, with its voracious appetite for content. To bolster its chances, Martin Sheen (Charlie's father) has joined the cast on a semiregular basis, giving the series, in the words of FX's CEO John Landgraf, "an extra dimension and mak[ing] it a multi-generational family show" ("Charlie Sheen Happy"). So far, *Anger Management* is attracting acceptable ratings numbers, and with this new model, conventional TV meets the web. The main idea here is product pumped out on a continual, reliable, workmanlike basis.

Quality is another matter. The production schedule of *Anger Management* gives new meaning to the old television director's motto "Get it, then forget it." As the series' executive producer put it, "The actors get the lines, we see the scene, the writers make changes, the actors go to makeup, cameras are blocked, we come back together and shoot the scene," and that's it ("Charlie Sheen Happy"). Whatever they get is good enough, and they use it. It's immediately created, instantly disposable, and then disseminated at lightning speed. Where will the syndication episodes wind up? Perhaps they'll be broadcast on television, but more likely, they'll be streamed on Hulu or a similar web platform. That's where the world is headed.

So streaming video, in the final analysis, offers us the past, the present, and the future. Which world will we choose? What kinds of audiences will view this plethora of programming? Who is out there on the other side of the screen? And what happens to conventional theatrical presentation in the new streaming universe?

2

The Lost Age of Classicism

On June 30, 2011, Chuck Viane, then president of Walt Disney Studios' Motion Pictures Division, wrote an open letter to exhibitors, shortly before his retirement. The point of the letter was simple and direct: 35mm is becoming obsolete; adapt to digital, or face obsolescence. As Viane wrote:

> At the end of this month I conclude 40 years in exhibition and distribution. I do not want to make this transition without sharing my deep concern with you. Some of you are among the dwindling number still playing only 35mm prints, apparently without plans to migrate to digital cinema. . . . The window of opportunity is closing for you to take advantage of our VPF [virtual print fee; more on this later] contributions to convert to digital cinema. At the same time, 35mm print costs are rising as suppliers grapple with falling volumes and soaring input costs such as silver and oil. (Film stock costs are up about 20% in just the past year.) I can't predict when, but we may reach a point when it is no longer economic for us to supply film prints on the same terms we have in the past, or at all. Likewise, it may become uneconomic for our suppliers to remain in

the 35mm print business. Under these circumstances, if you intend to remain a long term player in the theatrical exhibition business, why take the risk of the eroding economics and questionable prospects for 35mm? The question is particularly timely and urgent when you can still take advantage of limited-time distributor-subsidized programs to convert to digital cinema. I urge you, as a friend in exhibition, not to miss the best opportunity you will ever have to upgrade. For the sake of your business and the moviegoers that we serve, I urge you make the move to digital, and do it now.

Or, as John Fithian, head of the National Association of Theatre Owners (NATO), has repeatedly said in numerous public addresses to theater owners and operators, "Convert or die."

Needless to say, Viane's letter and Fithian's exhortation aren't suggestions: they're ultimatums. Just as in 1927, when talkies wiped out silent film production, digital cinema has sounded the death knell for conventional 35mm film. Mike Hurley, who owns two theaters in Maine, knows precisely whereof he speaks:

Many theaters that never thought they'd go digital are now adopting at a fast pace. One of my theaters, The Colonial Theatre, will be 100 years old in April. We're in the midst of conversion; I accept and embrace that day. Every time I see platter scratches, or receive a scratched and dirty print, or deal with a particularly odd projectionist, I look forward to it more and more. But it hasn't happened fast enough. At the end of 2011, Fox announced they'd no longer release product in 35mm "sometime in the next year or two." Also ending soon: The VPF, or virtual print fee. Since 2009, film distributors have paid VPFs to exhibitors. Based on the difference between the cost of a celluloid print and digital delivery, it's designed to help theater owners offset the cost of a digital cinema retrofit, which costs about $65,000 at the low end. (A new projector, by comparison, was about $20,000—but that was before you'd pay people to take them away.) The VPF has helped some, but not all. As a result, NATO recently estimated that up to 20% of theaters in North America, representing up to 10,000 screens, would not convert and would probably close. "Convert or die," indeed. And that's from someone representing theater owners.

But in reality, there is no choice. As Hurley notes, 35mm prints are going to cease to exist, and 35mm raw film stock is increasingly hard to come by; nearly everything today is shot in digital format. Theaters that don't convert will simply have no product to run, except for older, archival films. That would be great, but most people simply stream these classic titles or buy or rent the DVDs.

In the *Los Angeles Times*, Mark Olsen offered a fascinating glimpse into the differences between digital cinematography and working with conventional 35mm film, as discussed by some people who know what they're talking about: the 2012 Oscar nominees for cinematography. Olsen writes:

This year's Oscar nominees for cinematography present a particularly varied cross-section of contemporary filmmaking at a time when the very infrastructure of how movies are made and seen is in transition. Consider: 35-millimeter film prints are being phased out in favor of digital projection. Consumer still cameras can be used to shoot high-definition digital video. Video on demand is becoming a popular viewing option. Even the venerable Eastman Kodak, which produces the film stock on which many movies are made, recently filed for bankruptcy protection. The Scandinavian-modern *The Girl with the Dragon Tattoo* was shot with digital cameras; the World War I–set *War Horse* was shot on film. *Hugo* was shot in digital 3-D to portray 1931 Paris, while *The Artist* was shot on color film, then transferred to black-and-white to evoke the end of the silent film era in Hollywood. *The Tree of Life* used footage shot both on film and digital and integrates nature photography into its storytelling.

And what's driving the entire process—as always, with the cinema—is money. As Gendy Alimurung notes:

> It costs about $1,500 to print one copy of a movie on 35mm film and ship it to theaters in its heavy metal canister. Multiply that by 4,000 copies—one for each movie on each screen in each multiplex around the country—and the numbers start to get ugly. By comparison, putting out a digital copy costs a mere $150. "Distributing movies digitally into theaters has been the holy grail of the studios," former Universal Pictures chairman Tom Pollock told *Variety* back in 2010. "They stand to eliminate billions of dollars in costs in coming years without spending very much." [In 2012], for the first time in history, celluloid ceased to be the world's prevailing movie-projector technology. By the end of 2012, according to IHS Screen Digest Cinema Intelligence Service, the majority of theaters will be showing movies digitally. By 2013, film will slip to niche status, shown in only a third of theaters. By 2015, used in a paltry 17 percent of global cinemas, venerable old 35mm film will be mostly gone.

Revival houses are now having trouble locating 35mm prints of classic films for exhibition; even if a studio has a 35mm print of an older title, it is now increasingly reluctant to rent it to theaters. Want to run *Breakfast at Tiffany's*? Pick either a digital cinema package (DCP) or a conventional DVD. That's it—35mm isn't an option. In November 2011, 20th Century Fox added its voice to the digital changeover, announcing to exhibitors that "the date is fast approaching when 20th Century Fox and Fox Searchlight will adopt the digital format as the *only* format in which it will theatrically distribute its films" (as quoted in Alimurung; emphasis added). Fithian's "convert or die" comment seems—for better or worse—simply common sense. As he put it at the 2011 NATO Convention in Las Vegas, "If you don't make the decision to get on the digital train soon, you will be making the decision to get out of the business" (as quoted in Alimurung). In short, there is no choice.

Projectionists also have to adapt to the demands of new technology, and while some embrace the changeover, others are not so sanguine. As Vinny Jefchak, a projectionist at the New Beverly Theatre, put it, his skills are now essentially obsolescent. Alimurung notes:

Playing a movie on a DCP projector involves plugging the hard drive into the projector, creating a playlist, as you would on an iPod, and pressing a button to play. "You could train a monkey to do it," Jefchak says. "Now they need to corner the market on monkeys." Jefchak works at the New Beverly, which is owned by Quentin Tarantino. A regular at the art-house cinema, Tarantino bought the place in 2007, when it was in danger of closing. The New Beverly still plays traditional reel-to-reel 35mm, and Tarantino has said that the day the cinema puts in a digital projector is the day he burns it to the ground. Recalling the quote, Jefchak laughs. "I don't know how to break it to him, but we've been running digital here for as long as we've had video projectors. But I think what he's trying to say is if we go exclusively digital because there's no 35mm print, then he will feel there's no reason to own this place anymore."

And naturally, the processing laboratories that make the prints of theatrical motion pictures are also facing a slow death. Until 2012, the major studios spent an average of $850 million a year on prints, and $450 million on shipping and handling, for a standard run of 3,000 prints for "saturation booking"—that is, the simultaneous opening in every available theater worldwide, to avoid piracy and increase box-office revenue before any possible negative word of mouth sets in (see Alimurung). That's all gone now. In 2011 longtime rival labs Technicolor and Deluxe merged, simply to stay in business (Alimurung). But how long can they survive when all demand for film copies of motion pictures has vanished?

The problems don't end there. As noted earlier, it costs substantially more to preserve a "born digital" movie than one shot on conventional 35mm film. Worse still, as digital formats change, the movie has to be migrated to each new system. Even relatively recent films can vanish without a trace; both *Toy Story* (1995) and *Toy Story 2* (1999) were nearly lost to oblivion because of digital format shifts and data corruption. As Alimurung recounts:

Five years after the first *Toy Story* came out, producers wanted to release it on DVD. When they went back to the original animation files, they realized that 20 percent of the data had been corrupted and was now unusable. [Backup copies on conventional film filled the gaps.]

Fast-forward to *Toy Story 2*, which was almost erased from history. Pixar stored the *Toy Story 2* files on a Linux machine. One afternoon, someone accidentally hit the delete key sequence on the drive. The movie started disappearing. First Woody's hat went. Then his boots. Then his body. Then entire scenes. Imagine the horror: 20 people's work for two years, erased in 20 seconds. Animators were able to reconstitute the missing elements purely by chance: Pixar's visual arts director had just had a baby, and she'd brought a copy of the movie—the only remaining copy—with her to work on at home. In the digital realm, the archivist's mantra, "Store and ignore," fails. If you don't "refresh," or occasionally turn on a hard drive, it stops working. You can't just stick it on a shelf and forget about it. As restoration expert Ross Lipman says, "You're shifting from a model focused on a physical object to data. And where the data lives will be constantly changing."

These were both decidedly mainstream films. If Pixar can screw up *Toy Story*, what chance does a small-scale digital film, perhaps an "orphan" film, have, with no one to look after it?

On film, a copy might last several lifetimes; on digital, only a few years. But as with all things digital, it is the initial ease of projection, production, editing, and mixing that attracts both studios and filmmakers; they're not nearly as concerned about the long view. At CinemaCon, the annual NATO conference held in April 2012, John Fithian again repeated his mantra of the inevitability of an all-digital motion picture landscape: "Last year, I predicted that domestic distribution of movies in the format of celluloid film could cease by the end of 2013. That prediction is becoming a reality" (as quoted in Bowles). As of April 2012, of the 40,000 theater screens in the United States, 27,000 were digital, with 1,000 more converting to digital each month (Bowles). Fithian is probably right; by the end of 2013, about the time this book appears, 35mm film will be only a memory.

Christopher Nolan is arguably one of the most interesting and innovative commercial directors now working in the cinema, with films such as *Following* (1999) and *The Prestige* (2006) to his credit. Remarkably, Nolan remains resolutely opposed to digital production, but what is even more remarkable is that whenever he does publicity for one of his new films, he spends more time arguing against digital cinema than plugging the movie he's spent years making. As recently as June 9, 2012, just weeks before *The Dark Knight Rises* opened in theaters around the world, Nolan took advantage of a press screening to state categorically, "There's a huge danger in all of this. . . . If you are looking strictly at production cost, then you would use digital. But for the best image, it is still film." As Alex Ben Block reported:

Nolan said that moving to digital creates a risk of "devaluing what we do as filmmakers." "The problem with the push to digital is [it] has been given a consumer aspect," says Nolan, who suggests it confuses the camera with an iPad. "It's not what is best for the film," he insists. While digital has made great strides, Nolan believes it has a ways to go yet before it will offer the quality to capture images that film does. "I don't want to be the R and D department. I don't have any interest in the research into electronics. What interests me is to use the best technology and that is film." Nolan says he does use digital technology in the editing process and for special effects and in other ways, but ultimately he wants his movies shot on film and shown on film. When the digital technology evolves to the point it has the same depth, image quality and look as film, he is open to shifting his view. "When it is as good as film and it makes sense I'll be open to it," says Nolan. "But (at present) it's not good enough."

With digital production, you can erase whatever you don't want from the camera's hard drive while shooting, but you also run the risk of jettisoning material that may prove useful later on. Nolan uses very few takes for each shot in his films, never employs a second unit director for the action sequences, and functions in general like a Howard Hawks, the great Hollywood producer-director from the medium's golden era who personally oversaw every aspect of the numerous films he created during his long career.

David Fincher, in contrast, used more than ninety takes for a single sequence in *The Social Network* (2010) and constantly erases material on the set before he has a chance to view it in the editing room. How can one possibly know what values might be lost in the process? The very ephemerality of digital cinema, which some see as its chief asset, is also a key vulnerability: once a take is erased, it's gone forever. With film, the choices are fixed, present, available. William Wyler, another luminary of the Hollywood studio era, was known to do repeated takes—sometimes as many as twenty, thirty, or forty—to get a scene right; he would let the actors relax and drop their mannerisms and stage tricks, and allow the reality of the sequence to come out. But after viewing the material on the dailies, Wyler had the option of using any take—they were all there because with film, everything is kept. Not so with digital cinema; if you don't make a conscious decision to preserve it, it's gone.

Cinematographers are also divided about the use of digital cameras in their work, as previously mentioned. During an interview with Mark Olsen, a widely disparate group of cinematographers discussed the increasing pervasiveness of digital cinema, but only one, Janusz Kaminski, came out forcefully against it. According to Kaminski, digital cinema represents "the death of the cinematographer . . . generally speaking, I don't have respect for digital media just yet. If you see the image on the digital screen I think people will become lazy, they get satisfied with just seeing the image, they're not going for visual panache, not getting the story through metaphors. With film there is still mystery" (as quoted in Olsen). Since Kaminski is a two-time Academy Award winner for Best Cinematography—for Steven Spielberg's *Schindler's List* (1993), which was shot, for the most part, in black and white, and for *Saving Private Ryan* (1998)—his opinion carries some weight.

But Kaminski was in the distinct minority; most of the other cinematographers interviewed were perfectly comfortable in either medium. Jeff Cronenweth, who shot David Fincher's 2011 remake of *The Girl with the Dragon Tattoo*, was an old hand at digital, having shot *The Social Network* with Fincher the year before. But Cronenweth made an interesting point as the discussion was drawing to a close. As time goes on, "it's going to be less of a debate," he said. "In all fairness, we're at the infancy stage of digital cinema" (as quoted in Olsen). And this is certainly true. When movies were first invented, they were photographed on paper film, which gave way to cellulose nitrate film stock and then, around 1950, to safety film. Digital is simply the next step. It comes with its own built-in set of advantages and drawbacks, but like sound and color, it is a technology whose time has come. And cinema is, above all, a medium driven by technology. As Mike Hurley observes:

This isn't the first time technological evolution has hit the film and exhibition industry, but in the past the development of new equipment was steady, orderly—and slower. That meant as early adopters grabbed the latest contraptions, there was a healthy market in used equipment for the smaller and less-profitable theaters. However, the brain trust in Hollywood seem[s] committed to playing a game of diminishing exhibition returns and appears ready to write off huge swaths of the ticket-buying public. You can bet that the same people who spent $150 million to make *Mars Needs Moms* have crunched the numbers and believe they can live with a lot fewer theaters in this world.

These theaters will also screen a much smaller selection of films, sort of a "top forty" format that will leave small productions behind. Even though more "independent" (read "low-budget") films are being made, they are having a much harder time finding a theater screen and usually end up on DVD, as a streaming download, or as video on demand (VOD). As Michael Cieply noted in April 2012:

Over the last 10 years, there has been a 74 percent increase in the number of movies being distributed by companies other than members of the M.P.A.A.—to 469 in 2011, from 270 in 2002. That happened even as the film count from those big member companies—Fox, Disney, Sony, Warner, Paramount, and Universal—and their units fell 31 percent, to 141 from 205 over those same years. But a closer look shows how the nature of an independent release has changed as well, and not necessarily in a way that puts those movies in front of more theatrical ticket buyers. The figures showed a near collapse in the number of films being released by the studio specialty units, as Disney sold Miramax and Paramount diminished its small-movie presence. In all, the number of films from the subsidiaries, which had been relatively well heeled and often gave their pictures a fairly wide release, fell 55 percent, to 37 in 2011 from 82 in 2002.

It's mainstream movies or nothing at all. Just look at the offerings at your local multiplex (there are few single-screen theaters left—it simply doesn't make economic sense). Perhaps, if you're lucky, you have a small "art house" theater in your neighborhood. If not, you'll notice that all the theaters in your area—indeed, around the country and throughout most of the Western world (England, France, Germany, Holland, Belgium)—are playing the same major releases, all backed by huge ad campaigns, both print and streaming video, in addition to appearances by the films' stars on Letterman, Leno, and the other talk shows on the public relations circuit. All this costs money, and the studios aren't about to put significant advertising dollars behind something that isn't presold and pretested—an almost guaranteed hit. Even if the film in question is terrible, trailer production companies like Buddha Jones or Workshop Creative can chop some scenes out of it to make it seem compelling for three minutes or so.

In January 2012 I participated in an NPR piece on the evolution of the motion picture trailer, with moderator Brent Baughman. I pointed out that, once upon a time, trailers—or movie advertisements showing "coming attractions"—relied mostly on superlatives and hyperbole to get the job done. As the transcript of the broadcast documents:

[Early trailers were] often full of over-the-top superlatives, like the trailer for *Gone with the Wind*. "Never so tremendous!" Dixon says by way of example. "The screen's greatest achievement!" One critic at the time said it was the supreme example of writing so as never to be believed. Compare that with something like last year's trailer for *The Dark Knight Rises*, which set a record for downloads in 2011. "The shots are shorter and shorter and shorter, and more fragmented," Dixon says. "There have been a number of studies that demonstrate that the average lengths of a shot in a film have been shrinking every single year, because audiences absorb information faster—and there's also a sense that you don't want to bore them."

Yet one might argue that contemporary trailers are all so similar that they wind up being peculiarly ineffective. Everyone seems to realize, even subliminally, that they're being lied to—that they're being shown only the most eye-catching moments in a desperate attempt to lure an audience that has been specifically targeted long in advance through the use of focus groups and test screenings of rough cuts of the film. And, as Baughman notes, even after a "successful" trailer has been created to market a major film, it's still subject to repeated revisions by other hands, including the grafting of sections of competing trailers into the final version. According to Baughman:

Studios take a whatever-sticks approach—sometimes assigning TV spots to one house and Internet trailers to another. Sometimes a producer might see his trailer in a theater, cut together with someone else's work. All that—so you can turn to your seat-mate in that green glow before the next trailer starts and murmur, "Yeah. I guess I'd see that." "They're a completely different art form," Dixon says. "They're an advertising art form, but they want to emotionally involve you." [The trailer editors offer] no argument. "Oh, yeah," [says one], "we're trying to seduce you. On our best days, we do that."

It all sounds a bit like outright deception, and that's just what it is. But this same sort of "packaging reality" or, in this case, "constructed reality" now pervades American society and social discourse as a whole, to the point that any essence of the real is so mediated that verisimilitude is lost. Even George Lucas, whose life's work has been fantasy films, is disenchanted with the way real life is skewed so heavily toward rampant consumerism; there's a pervasive sense of disillusionment, as if we collectively know we're being deceived. As Lucas told Charlie Rose:

> Movie-making is soaring, because we've developed digital technology. The equipment is smaller. It's cheaper. And it's now becoming democratized, so anybody can make a movie. [But] what that means to the studios is a whole other issue, because they're now caught in a transitional period that they don't really know what to do with. They don't know how to adapt. They are corporations where there's layers and layers and layers of executives that don't know anything about making movies. And they have lost, which is what happens in that corporate environment, any respect for the workers that actually do the work. They don't realize how hard it is or what they do or the value of the worker. And as a result, they just think that you can go out and, in my case, anybody can direct a movie. Anybody can write a movie. Just put anybody in there. It doesn't make any difference.

Of course, it *does* make a difference. And whether you're working on the Hollywood model or the lowest-budget independent model, what makes the difference in any film is the individual vision of the filmmaker. He or she is the ultimate arbiter of what goes into the film, what values it holds dear. More than anything else, the quality of the work itself dictates whether a film will transcend time, surviving changing production plat-forms and passing fashion to create a lasting impression with each successive generation. Nowhere was this truer than in the 16mm experimental film movement of the 1960s, often called the "underground film" period. This was when some of the most explosive visions in American cinema history were created, and surprisingly, it was when George Lucas began his career. I re-member seeing Lucas's student film *THX 1138*—later expanded into a feature—at the Yale Film Festival in the late 1960s. It was one of a group of films, made by widely divergent artists, ad-dressing political, social, and sexual concerns. In short, the best films are always personal, the work of one individual alone, and the 16mm generation of the 1960s paved the way for today's low-budget, independent digital artists.

In the 1960s, while working in New York, I was part of a group of filmmakers who created films out of almost nothing at all—outdated raw stock, ancient cameras that barely functioned and were often borrowed from someone else, a few lights, the barest outline of a script, and "financing" that consisted of donat-ed labor both in front of and behind the camera. Nobody had any money; we lived in cheap apartments that cost as little as $100 a month, worked a variety of odd jobs to keep the wolf from the door, and plowed nearly everything we made back into our films. These films had no market, no commercial value, and were so resolutely personal that it seemed that no one, outside of a small circle of friends, could ever find them of value, worth, or interest.

Sync-sound filmmaking equipment, only recently invented at that point, was beyond our financial range; so, like the early silent filmmakers, we were forced to depend on the primacy of the image and created films of deeply romantic intent using a few costumes, borrowed props, and the barest of sets. Another defining characteristic of these films was their calculated sloppiness, since we were dealing with second-, third-, and fourth-rate equipment and film of uncertain origin. But it was all we could afford. So we used every possible frame of what we shot, down to the last bit of leader-streaked material at the end of the roll, in a desperate attempt to capture every last bit of our vision on film.

Shoots were organized in a very informal manner; for Gerard Malanga's film *In Search of the Miraculous* (1966), for example, Gerard and the late Warren Sonbert (a gifted filmmaker in his own right) simply went to the Columbia University library early one morning and staged an impromptu dance sequence (which opens the film) in a matter of minutes. Warren photographed the scene while walking behind the pillars of the library, so the viewer's vision is obscured at regular intervals throughout the three-minute sequence. Every last frame of the footage was used, and the shoot itself was completed in fifteen minutes or less.

The structure of these films—which Man Ray would describe as "cinepoems"—often consisted of simply shooting material that seemed to circulate around a certain theme or subject. Then, when a sort of creative critical mass had been achieved, we would rent or borrow editing equipment for a marathon weekend editing session in which the material would be integrated in a fashion dictated by equal amounts of chance, intuition, and thematic resonance. If the finished film was to have a soundtrack, it was usually compiled from records—often scratched to death. The music would be mixed and remixed on ancient reel-to-reel tape recorders; then, if one could arrange it, it would be transferred to an optical soundtrack in a clandestine "midnight session" at one of the city's many sound transfer locations, so that the final print would have a proper soundtrack for screening at festivals.

In the mid-1960s there were perhaps 400 people in all of Manhattan engaged in such work, defiantly existing in the most poverty-stricken circumstances but also making a conscious artistic decision to reject the Hollywood model of scripts, stars, narrative cohesion, and gloss for a rougher, tougher, more resolutely stripped down vision of life. Peter Emanuel Goldman, for example, created his superb film *Echoes of Silence* (1965) out of nothing more than a large supply of outdated 16mm film, a spring-wound Bolex, and existing sets and locations, using his friends as actors. After starting the film and then discarding that material, Goldman realized that by simply keeping as close as possible to the actors' faces—more or less setting them in a situation and observing them—an essential reality came out in the material that had otherwise eluded him. Shot for a total of $1,600 to final print, the finished film was screened at the New York Film Festival in 1966.

In San Francisco, Ron Rice shot *The Flower Thief* (1960) on outdated World War II aerial gunnery film donated by, of all people, the legendarily cost-conscious Hollywood producer Sam Katzman. Working with the sublimely talented and Chaplinesque Taylor Mead, Rice completed an entirely improvised seventy-five-minute film on a budget of less than $1,000, with the soundtrack comprising—once again—snatches of classical and pop music from old, scratched-up records interspersed with segments of Beat poetry. As Rice said of the film in the program notes for the premiere, "In the old Hollywood days movie studios would keep a man on the set who, when all other sources of ideas failed (writers, directors) was called upon to 'cook up' something for filming. He was called The Wild Man. *The Flower Thief* has been put together in memory of all dead wild men who died unnoticed in the field of stunt."

There are numerous other examples, both well known and obscure. Andy Warhol shot his seventy-minute take on Anthony Burgess's novel *A Clockwork Orange*—rechristened *Vinyl* (1965), with a scenario by Ronal Tavel—in a single afternoon in 16mm sync-sound for a total budget of roughly $300, using donated labor and performers who worked simply because they believed in the project. Robert Nelson shot his memorable attack on racism, *Oh Dem Watermelons* (1965), using the services of the San Francisco Mime Troupe, with music composed and performed by Steve Reich, also for just a few hundred dollars. Vernon Zimmerman's astoundingly ambitious twenty-six-minute film *Lemon Hearts* (1960), featuring Taylor Mead playing no fewer than eleven roles, was shot for a total cost of $50 to final print, using every possible means of economy and barter. It offers a poetic, tragicomic vision of the absurdity of modern life, with Mead wandering through the ruins of some demolished Victorian houses while musing on the soundtrack about the heartlessness and futility of "civilization."

None of these films had an audience or any hope of making even their minimal production costs back; they were produced because the filmmakers wanted to create them and found congenial, talented colleagues who were willing to participate in what was then viewed as utterly "worthless" work. Today, all the films mentioned above are acknowledged classics of the avant-garde, and if they have retained an air of invisibility over the years because they never made the jump to DVD and, in many cases, are "orphan" films with no one to care for the originals, they still exist as difficult-to-see prints in archives and museums, authentic talismans of the era. Ironically, they have attained a degree of rarity and value simply because their vision of the world is at once so raw and innocent that it authentically recalls a time that is almost incomprehensible to the contemporary viewer. Created out of nothing, these films now hold the poignant vision of a vanished culture in which money was a means to an end and inherently suspect, because the moment significant capital was invested in a project, it almost inevitably became compromised.

Make no mistake about it: each of these films, though created by loosely thrown together teams of actors and technicians, was the result of one person's concept. As rough-hewn as they were, and despite the considerable budgetary limitations, the filmmaker's vision alone was the driving, informing voice of the work. Today, with the easy availability of video production, nearly everyone has tried his or her hand at making a "film" of one sort or another, but these projects are often parodies of other films or genre pieces informed by the history of film itself; they reside within the boundaries of conventional narrative, rather than coming from a place entirely outside the existing culture. In contrast, the previously mentioned "outlaw" films were made by people who paid dearly for their work (Ron Rice, for example, died at twenty-nine in Mexico City of illness and malnutrition because he spent every penny he had on his films, leaving nothing for his own existence—something akin to the equally tragic example of 1930s French filmmaker Jean Vigo).

Today, Manhattan has been gentrified beyond all recognition. There are no cheap apartments; no places to hang out and drink coffee all day because you have no money for food; no cheap loft space in which to create raw, powerful work out of scrap material; no Lower East Side "Alphabet City" of mean streets; and little or no community dedicated to the pursuit of supposedly "worthless" work. Even the outer boroughs are ridiculously expensive. It used to be, in Thomas (not Tom) Wolfe's words, that "only the dead know Brooklyn." Now, it too has become a tourist-friendly recreation zone.

Andy Warhol found it hard to sell his pop paintings in the early 1960s because they seemed so "easy" to make. They could be silk-screened in a matter of seconds, and it was assumed that Warhol would always be there to make another painting and then another. Indeed, in 1962 or so, one could buy a Warhol painting for as little as $100 if one went directly to the artist's studio, located in a rundown firehouse with no heat, water, or electricity. And of course, Vincent van Gogh didn't sell a single painting in his entire lifetime, despite the fact that his brother Theo was an art dealer; today, van Gogh is an industry. Henry Miller spent much of his life working in a cabin, writing books he thought no one would ever publish, living on the barest necessities, and cadging loans from Frances Steloff of the Gotham Book Mart. For many years, his pleading letters for $20 or $40 were stuck up on a bulletin board in the front of the store, mute testimony to Miller's precarious existence.

Today, it seems that the only art is making money. There are plaques all over New York City identifying where this artist or that artist used to have a studio; all the locations are now office buildings or banks. Given the current financial crisis, it seems that no one has time or money for artistic work. But in fact, such work would redeem us as a society, as it did in the 1930s when Franklin Roosevelt put artists to work and then sold their efforts to get that segment of the economy moving again. Now, the stranglehold of social conservatism that pervades the nation belatedly recognizes the power of "outlaw" art and no longer wishes to support it because it might well prove dangerous in the long run.

Money can create, but it can also destroy. Out of economic privation and the desperate need to create, the aforementioned artists produced works of lasting resonance and beauty with almost no resources at their disposal, other than the goodwill and assistance of their colleagues—a band of artistic outlaws. These artists broke the mold of stylistic representation in the cinema and offered something new, brutal, and unvarnished that confronted audiences with a different kind of beauty—the beauty of the outsider, gesturing toward that which holds real worth in any society that prizes artistic endeavor. Only work that comes from the margins has any real, lasting value; institutional art, created for a price or on commission, documents the powerful and influential but doesn't point in a new direction. Work that operates off the grid and under the most extreme conditions, without hype or self-promotion, has the greatest lasting value, precisely because it was created under such difficult circumstances.

None of these films, however, has made the jump to digital; they reside in 16mm film cans in independent distributorships. Ostensibly, they are available for rent, but by whom? No one has 16mm projectors anymore; only a few archives around the United States or, more accurately, anywhere in the world have the ability to project 16mm films. Once, these films were screened ubiquitously, when every college film program relied on the 16mm format. Now they're virtually invisible, as if they never existed. An entire era, as well as a system of cultural values, has been rendered obsolete. When the times change, we must change with them; there's too much to lose otherwise. Many of these films are "orphans" whose makers have died or moved on to other fields; in some cases, even the filmmakers themselves have forgotten about the films, which seems the ultimate irony.

Even mainstream Hollywood filmmaking from the 1920s through 1980s is commercially suspect; that's why it's so important to have Turner Classic Movies (TCM), a 24-hour streaming movie channel that runs its entire library, from shorts to feature films, on cable television, available to a wide audience. But how much longer can it hold on in such a hypercommercial environment? Indeed, has there ever been a more ambitious, more altruistic project in the history of television than TCM's showing of commercial-free, edit-free movies screened in the proper aspect ratio? I don't think so. It's a nonstop film festival at home. Of course, TCM needs and wants to make money, but money doesn't seem to be the bottom line here; it's the history of cinema that TCM celebrates 24 hours a day, 7 days a week, 365 days a year. This is an astounding accomplishment. Through its continually changing programming, as well as its amazingly complete website, TCM offers a crash course in films for the uninitiated, and for those of us who grew up with the movies, it offers an invaluable alternative to the avalanche of dreck that passes for contemporary television programming.

AMC (formerly known as American Movie Classics) once ran all its films commercial free, but now they are slashed to ribbons, with hundreds of commercials; the same is true of the Independent Film Channel (IFC). Both AMC and IFC can accurately state that they don't cut the movies they run; the films are complete, but they're constantly intercut with advertisements, staggered in the usual manner of fewer at the top, so the viewer can become engrossed in the film, and more at the end, after the viewer has been properly hooked. TCM, thus far, has managed to avoid this and continues to serve up an incredible variety of films from around the globe, with host Robert Osborne guiding us through film history. In an era when such exploitational trash as *1,000 Ways to Die* passes for "entertainment," it's nice to have a refuge from the ceaseless commercialism that cable television was supposed to deliver us from. This is what television was intended to be at its inception: too bad it turned into a commercial wrapped in a promo hidden inside an elaborate product placement.

When American Movie Classics first came on the air, it had a half-day schedule, splitting its satellite time with another network, and a somewhat limited playlist. Nevertheless, all the films it ran were uncut and commercial free and were presented in their original aspect ratios, whether Academy, widescreen, or CinemaScope (and their related formats). Eventually, American Movie Classics became a twenty-four-hour network, still running commercial-free, uncut classic films. Then it "rebranded" itself as AMC and began intercutting its films (still shown in their entirety) with hundreds of commercials, completely ruining the films' impact. Likewise, IFC presented films uncut and commercial free for many years but recently began running ads—while still advertising the films as "uncut"—forcing you to watch commercials for this product or that. This change might have been an attempt to entice consumers to buy the IFC in Theaters service, which I use quite frequently: first-run films (uncut, commercial free, and in their original aspect ratios) are presented on cable for a per-film fee on the same day they open in theaters in "selected cities." It's cheaper than going to the theater, especially if the nearest one running the film is 1,000 miles away.

Now there are only two basic cable services left that run feature films uncut and commercial free: the Sundance Channel, which has a somewhat limited catalog, and TCM. Importantly, TCM and Sundance nearly always show films in their original aspect ratios. If it was shot in CinemaScope, you see it in CinemaScope, with the signature black bars at the top and bottom of the screen; if shot in widescreen, there are slightly smaller bars; if shot in Academy ratio, the film is shown in full frame. This is something you can't say of HBO, Showtime, or the other so-called premium channels, which, as a rule, show "pan and scan" versions of CinemaScope and widescreen films. This means that up to half the original image is lost, all in the name of filling up the entire screen, even if it's only half of what the director photographed. Martin Scorsese has called pan and scan tantamount to "redirecting the movie"—the sides of the frame are cut off, backgrounds are eliminated, and characters are chopped out of the frame. With pan and scan, you get a "full frame," with no black bars at the top and bottom, but you're not seeing the whole film. You get less, not more. HBO and other premium channels offer what they term "wide" versions of some films for their on-demand customers, but for their regular offerings, pan and scan is the rule.

To summarize, movies are shown with copious commercials on IFC, AMC, and all the other basic cable channels; "pan and scan" versions run on HBO, Cinemax, and the other premium cable services. So if you want to watch feature films as their makers intended them to be viewed—in their original aspect ratios and without commercials, time compression, or editing—you have only two choices: TCM and Sundance.

But even then, it isn't the same experience. Although it's fine that digital copies of the masterpieces of the past (and the junk, for that matter) are available for viewing, it would be nice to have a choice in the matter. That's something that has been taken away from us—and apparently, we didn't even notice. Recently, director Christopher Nolan came out forcibly in favor of film as an originating format. As he points out in a *DGA Quarterly* interview with Jeffrey Ressner:

> For the last 10 years, I've felt increasing pressure to stop shooting film and start shooting video, but I've never understood why. It's cheaper to work on film, it's far better looking, it's the technology that's been known and understood for a hundred years, and it's extremely reliable. I think, truthfully, it boils down to the economic interest of manufacturers and [a production] industry that makes more money through change rather than through maintaining the status quo. We save a lot of money shooting on film and projecting film and not doing digital intermediates. In fact, I've never done a digital intermediate. Photochemically, you can time film with a good timer in three or four passes, which takes about 12 to 14 hours as opposed to seven or eight weeks in a DI suite. That's the way everyone was do-

ing it 10 years ago, and I've just carried on making films in the way that works best and waiting until there's a good reason to change. But I haven't seen that reason yet.

I've kept my mouth shut about this for a long time and it's fine that everyone has a choice, but for me the choice is in real danger of disappearing. So right before Christmas [2011] I brought some filmmakers together [a group that included Edgar Wright, Joe Dante, Michael Bay, and Bryan Singer, to name a few] and showed them the prologue for *The Dark Knight Rises* that we shot on IMAX film, then cut from the original negative and printed. I wanted to give them a chance to see the potential, because I think IMAX is the best film format that was ever invented. It's the gold standard and what any other technology has to match up to, but none have, in my opinion. The message I wanted to put out there was that no one is taking anyone's digital cameras away. But if we want film to continue as an option, and someone is working on a big studio movie with the resources and the power to insist [on] film, they should say so. I felt as if I didn't say anything, and then we started to lose that option, it would be a shame. When I look at a digitally acquired and projected image, it looks inferior against an original negative anamorphic print or an IMAX one.

But whether Nolan likes it or not, his films will probably be screened in their final form as DCPs, as more and more movie theaters convert to digital projection alone.

And because the theatrical audience is dwindling, what makes the most money are streaming downloads of films, once they've had a theatrical premiere to heighten their visibility in the marketplace. Cable television is also seen as increasingly old-fashioned; as Tony Cox of National Public Radio observed in a broadcast on August 2, 2011, "streaming video means primetime all the time." You can watch whatever and whenever you want to. As he put it:

> The days when everyone watched the same program at the same time are gone. DVR, Blu-ray, game consoles—all have made more and more people decide that maybe this is the year to cut the cable. Internet TV options are boundless and have even made piracy less of an issue. The days when your entertainment options were limited to the 300-plus channels on your cable box seem quaint nowadays. Netflix, Hulu, Amazon, HBO GO, Internet TV has made the vast wasteland truly vast, changing how and where we watch almost everything. Some people even ditched Netflix after its recent price hike, something that would have been unthinkable had there not been so many other alternatives.

Cox then observed that the embrace of digital technology and streaming might be a "generational thing." His guest on the broadcast, Farhad Manjoo of Slate.com, agreed, albeit with some qualifications. As Manjoo observed:

> I suspect it's a generational thing, but it's probably not a very distinct generational thing. I mean, obviously, there are people in college who are more . . . familiar with the Internet [and] are obviously . . . going to be at the forefront of this trend. I think that many older people who are looking at their huge cable bills every month might also consider this. . . . You can watch shows now on many different kinds of screens, and you can watch them kind of on the go. So you can watch movies and TV shows while you're on a plane or on a bus. And, you know, all of those things, I think, appeal to people . . . across all ages.

Manjoo then noted that, in the past, much of the traffic in streaming video was illegal—called "torrenting"—but illegal file sharing is now going out of favor, and Netflix and other legal content providers are gaining the upper hand. In short, an "underground" community has shifted to a resolutely commercial model. As Manjoo admitted, with some degree of guilt:

> A couple years ago, I confessed on Slate that I had been getting a lot of my entertainment illegally that way. But a couple weeks ago, I wrote that . . . I had stopped substantially using file sharing networks because there are now so many legal options that are so easy to use that—and cheap—that . . . the hassle of file sharing isn't worth it at all.
>
> And I think we're noticing this in the numbers. A couple months ago . . . an Internet traffic company reported that for the first time in North America, Internet traffic for Netflix had eclipsed Internet traffic for BitTorrent, which is the biggest [illegal] file sharing system for . . . TV shows and movies.

There are, in fact, plenty of free and legal options available on the web, the most conspicuous and squeaky-clean of which is the Internet Archive. It boasts a library of more than 3,000 feature films available for instant legal download, including *And Then There Were None, Scrooge, The Joe Louis Story, Night of the Living Dead*, the 1954 version of *The Fast and the Furious, The Inspector General, His Girl Friday, The Jungle Book* (the Korda version, of course), *The Lost World* (1925 version), Ida Lupino's *The Hitch-Hiker, The Big Combo*, and F. W. Murnau's *Nosferatu* (1922), as well as the deeply idiosyncratic films of German auteur Lutz Mommartz. And if that's not enough, the Internet Archive also has a remarkable collection of more than 11,365 classic television programs, commercials, government proceedings, and more—

all available at the touch of a button for either instant streaming or downloading. Some of the most popular collections are classic television commercials from the 1950s and 1960s and complete classic television programs such as *Dragnet*, *The Ed Sullivan Show*, *The Beverly Hillbillies*, and the BBC's production of George Orwell's *1984*. These and thousands of other programs are available on the site.

Music, books, and movies have all gone streaming. The convenience factor aside, it's easy to see that the rush to streaming creates many new ways to "monetize" back catalog images, music, and books in a way that perpetuates the consumer's dependence on the supplier and ensures a constant revenue stream not only to access but also to store, download, or view the works the consumer has supposedly "purchased." Once upon a time, about a half century ago, you bought a television, stuck an aerial on the roof, and got free programming, even though you had to put up with commercials interrupting your viewing. Now, you still get the commercials *and* you have to pay for the content, whether on cable or on the web.

And the old media are dying. *Variety*, the iconic show business newspaper that once dominated the field, has been replaced by the newer, sleeker, online-only *Deadline Hollywood*, founded by Nikki Finke. Twenty-four hours a day, seven days a week, Finke and her staff blog about the entertainment business, with a blizzard of posts that concentrate solely on the commercial aspects of the business—there's no aesthetics here, just numbers. Tellingly, *Deadline Hollywood* contains no film or television reviews, just statistics. It's all about whether something is commercially successful; opinions don't matter. If something makes money, that's a positive thing. If it doesn't, it's written off. These are the only yardsticks that apply, and *Deadline* is ruthless in its pursuit of the latest industry news: who is leaving what studio for another, more lucrative position; what agency has signed the "hottest" new talent; who's on the way up and who is falling out of favor—sort of like an ever-changing high school yearbook of who's hot and who's not.

In this new landscape, *Variety* simply seems out of place. *Variety* covers the financial aspects of show business but also features a massive review section, highlighting films from around the world and often "breaking" these films to American audiences, something that *Deadline Hollywood* simply isn't interested in. And with international filmmaking exploding and independent filmmaking on the rise exponentially, it would be nice to have an authoritative journal of record where these films could find an audience. After all, many of the Criterion Collection of classic feature films are available from Netflix as streaming downloads, proving that there's an audience for the classic foreign and experimental films of the past—they just have to find that audience. *Variety* once provided that platform, and it could do so again if it does three things: gets rid of the print edition and goes entirely digital; gets rid of the pay wall (the requirement that the reader must subscribe to view content), so it can compete with other web-based showbiz journals; and hires a group of dedicated twenty-somethings to run the paper—people who are willing to pound the keyboard 24/7 and use their copious industry contacts. I'd like to see *Variety* emerge from the ashes; I'd also like to see the review section restored to its former glory. An all-digital, twenty-four-hour rolling-deadline *Variety*—trading on the value of its name-brand recognition—is the only way the journal can come back.

When Sime Silverman founded *Variety* in 1905, he noted that the business was based on "constant change." This sentiment was echoed by MCA/Universal head Lew Wasserman many years later when he described the entertainment business as experiencing a rebirth with each new technological shift; the industry simply has to keep up with the times or get plowed under. The latest news on *Variety* is encouraging. In October 2012, Penske Media, the owner of *Deadline Hollywood*, bought *Variety* and announced that the journal will remain free on the web, as well as being available in print.

Miramax, which, for the most part, specialized in quality boutique movies, now streams all its back catalog through Netflix. Once owned by Disney, which purchased it from founders Harvey and Bob Weinstein, Miramax was spun off in 2010 for $660 million to Filmyard Holdings LLC, led by Los Angeles–based investors Ron Tutor and Tom Barrack. The main asset, of course, was Miramax's library of films, and with the Netflix deal, Miramax has moved confidently into the new streaming arena, carrying with it the values of an earlier, more classical model of filmmaking (see Fitzgerald). And there's a larger market for these films than one might think. As Netflix spokesperson Steve Swasey notes, "We found over time more and more Netflix members have gravitated to catalog, independent film, foreign films, and TV shows . . . [and] more than 70% of DVD rentals at Netflix [are] not new releases. . . . They might not be the hottest, latest new release, but a great movie from 1972 is still a great movie." According to George Feltenstein, senior vice president of theatrical catalog marketing at Warner Bros., in addition to streaming older titles, many consumers want them in a more permanent format on DVD. Although David Konow observes that, "with today's audience, access and instant gratification are a top priority," Feltenstein has discovered, through the Warner Archive's on-demand DVD service, that "people actually want to own physical product. People want to have a disc on the shelf with really nice packaging so they can own it. We're seeing our business strengthen in this era when people think there's less interest in physical media or older things" (as quoted in Konow).

Partly, this is because nobody knows what will be a hit and what won't; the business of film production, distribution, and exhibition is still highly speculative. This is what many people simply don't understand: the runaway hits and new franchises always come from the margins, not from films that were designed as commercial enterprises. The future of cinema comes from original ideas, not retreads of what has already been done. As Nancy Tartaglione reports in *Deadline Hollywood*:

> British Prime Minister David Cameron [visited] Pinewood studios [to urge] filmmakers to ramp up their efforts to rival Hollywood by making more commercially successful pictures [and suggested] the UK's Lottery funding scheme be rebalanced to support more mainstream films with commercial potential as well as culturally rewarding films. The news is likely to upset the independent film community with director Ken Loach already appearing on the BBC today to say: "This is a travesty. If you knew what was going to be successful before you made it then we'd all be millionaires. It doesn't work like that. Public money should go to fund a wide variety of projects and people."

Loach is absolutely on target. This sort of "bottom-line" thinking ignores the fact that *The King's Speech*, which Cameron held up as an example of a huge international hit produced in the United Kingdom, went through a torturous development period, was shot for practically nothing (the equivalent of US$15 million), and was never considered primarily a commercial project. It was the unique quality of the film, the superb performances and direction, and, of course, the marketing campaign that put it over the top. But bottom-line thinking always predominates.

When Matsushita Electronics bought MCA Universal in

the 1990s, a Matsushita executive seriously suggested at a "get-together" board meeting that Universal's bottom line would improve if the company produced "only hits." Needless to say, this is impossible. Britain's last major international hit maker was Hammer Films in the 1950s and 1960s, which revitalized horror movies. Today, the most interesting and successful British films (as in the days of channel 4 in the 1980s and the British New Wave of the early 1960s) continue to come from outsiders with new ideas who look at the present and see something new for the future—something the public hasn't been offered before.

Experimentation alone is the real future of the film industry, in the United Kingdom or anywhere else. Consider *The Devil Inside*. Shot for under $1 million in Bucharest, Romania, the film had raked in $35,128,858 in theatrical rentals as of January 9, 2012, despite terrible reviews and stiff competition at the box office. If offers a new slant on a very familiar theme, and people flocked to see it. Or, on a much higher plane, consider *The Hurt Locker, Margin Call, Capote, Slumdog Millionaire,* and *Melancholia,* to name just a few. All were made for very little money and because the people involved wanted to make these films, not because they were designed—if such a thing were possible—as hits. All were commercial and critical standouts. And all have significantly advanced the careers of their makers.

The newest ideas, the freshest inspirations, the strongest franchises don't come from the center or from "safe" bets—there's no such thing, of course. They come from people with a passion who have something new to offer audiences—something they haven't seen before, something they'll remember and tell their friends about, something they'll want to see again. This is also true of the classics. Many of the most interesting classic feature films and television shows are absolutely free on such sites as Crackle, Video Buzz, and GoFree.com. Likewise, Hulu Plus,

VUDU, and Blockbuster on Demand offer many foreign and independent films for minimal rental fees.

For all the commerciality of the Hollywood studio system, consider this: if it wanted to, Apple, a company that didn't even exist thirty years ago, could purchase all the major studios except for Disney with the staggering $100 billion in its cash reserve. For a business that is so influential on a worldwide basis, the actual profits generated by Hollywood are relatively small. In 2011 Paramount cleared just $1.98 billion; Warner Bros., $1.8 billion; Sony, $1.3 billion; Disney, $1.2 billion; Universal, $1 billion; and Fox, a mere $950 million (Finke). These aren't insignificant numbers, but if one were interested in sheer profit, sectors of the financial industry would certainly prove more lucrative.

The people involved in making movies—whether strictly mainstream or resolutely experimental—do so because they love it. Like cinematic pioneer Georges Méliès, celebrated in Martin Scorsese's *Hugo* (2011), they want to create a little bit of magic to lift audiences out of their everyday existences. If this sounds a bit too altruistic, it really isn't. The Marvel films, such as *The Avengers* (2012), are strictly driven by the desire for profit, as are Michael Bay's Transformer films, but at the same time, there are easier ways to make money if that is one's only motivation. For better or worse, depending on who is in charge, films represent a repository of both corporate and individual dreams, as well as cultural history, in which competing visions battle it out on the screen for our attention.

The most profitable films always aim the lowest. As Orson Welles once famously observed, "the best thing commercially, which is the worst artistically, by and large, is the most successful." That's as true today as it ever was. But everyone in the business of moving images—from television to streaming video to theatrical motion pictures and all the stops in between—is driven, in the final analysis, by the desire to build an imagistic world they can entirely control, by the urge to create their own universe, their own order of things. Otherwise, they'd simply join Goldman Sachs or Morgan Stanley. For all the attention to the bottom line, money isn't really the driving force here. It's a desire to put one's stamp on the world, to leave behind work that will entrance future viewers for as long as there is some system of moving image storage and retrieval. Cinematographer Jeff Cronenweth is right when he observes that digital cinema is in its infancy, but it will soon be the dominant standard, and the storage, distribution, and other problems will eventually be worked out.

Nevertheless, movie moguls today have much less power than they did in the days when the major studios, and then the three broadcast television networks, controlled the dissemination of moving images with a very tight grip. As Brooks Barnes noted on April 25, 2012:

> In the days of the great studio chiefs—Louis B. Mayer, Lew Wasserman—movie operations were still autonomous, allowing the executives in charge to wield absolute power. Even when consolidation swept deeper into Hollywood in the 1980s, studio chiefs still reigned supreme because, as [analyst Harold] Vogel put it, "the movies were still big enough parts of the puzzle that they could sway the stock prices and earnings of the conglomerates that were buying them." Today, studios are embedded so deeply in media and technology giants like Sony and Viacom that their activities barely register on Wall Street. Disney had total revenue of $41 billion last year, but only $6.4 billion came from the studio. When [Rich] Ross [former Disney CEO] . . . said last month that *John Carter* [2012], one of the movies made on his watch, would require a $200 million write-down,

the company's stock price remained stable. . . . Why run a movie studio when you can build the next YouTube? Add the movie industry's difficulties—a dying DVD business, shrinking attendance in North America, social media undermining traditional marketing techniques—and the jobs are even less desirable. [As Vogel summarizes], "there used to be a time when people would kill for these jobs, but running a studio is not what it used to be."

Streaming is one of the factors driving this shift. If you can simply order up a hit movie on your cable system or through the web, why would you patronize a theater? As has been thoroughly documented elsewhere, particularly in Gabriele Pedulla's polemic book *In Broad Daylight: Movies and Spectators after the Cinema*, most viewers are indifferent to the nuances of theatrical presentation; given the choice, they would rather stay home or watch even the most special effects–laden films on an iPad or other mobile device. As Pedulla notes, "The purism of the big screen's champions and their almost religious cult of the movie theater seem to have been vanquished by the common sense of the man on the street, who *has never wondered* whether [Federico Fellini's] *La Dolce Vita* [1960] on TV might be truly, deeply different from *La Dolce Vita* at the movies" (3; emphasis added). Pedulla's "has never wondered" argument is an essential component of the overall cinematic equation faced in the twenty-first century by both consumers and producers of moving image constructs: most viewers have never given the issue any thought, opting instead for the obvious ease and convenience of streaming video at home or on the go. Then there's the cost factor: as we've seen, streaming costs considerably less than a night out at the local multiplex.

But if viewers are embracing the shift to VOD streaming of movies on cable and the web, directors and producers have more definite ideas on the subject. They want to see the theatrical experience remain the dominant form of distribution and exhibition, if only because it lifts films above the flood of titles pouring through the various streaming video portals. As Pamela McClintock reported in April 2011, a powerful consortium of directors wrote an open letter opposing a relatively new phenomenon dubbed "premium VOD," which would put "movies from Warner Bros., Sony, Universal and 20th Century Fox in homes only 60 days after a film's theatrical release. Those signing the letter include Michael Bay, Kathryn Bigelow, Guillermo del Toro, Roland Emmerich, Peter Jackson, James Cameron, Shawn Levy, Michael Mann, Todd Phillips, Brett Ratner, Adam Shankman, Gore Verbinski and Robert Zemeckis." The letter asked in part that "our studio partners do not rashly undermine the current—and successful—system of releasing films in a sequential distribution window that encourages movie lovers to see films in the optimum, and most profitable, exhibition arena: the movie theaters of America."

As James Cameron told McClintock, "You can argue about VOD windows all day long, but what you can't deny is that there is an overwhelming outcry from the theater owners that they feel threatened by this. The cinema experience is the wellspring of our entire business, regardless of what platforms we trickle down to. If the exhibitors are worried, I'm worried. We should be listening to them. Why on earth would you give audiences an incentive to skip the highest and best form of your film?" This isn't just about big-budget spectaculars. Added maverick director Karyn Kusama, "As someone who hopes to have the ability to keep making small movies alongside the opportunity to make some bigger ones, I am concerned by how much a shortened VOD window might affect a filmmaker like me. This shortened window might imperil the robustness, and challenge the already shrinking flexibility in programming, of the very venue that makes movie-making, and movie-watching, the work we choose to do."

John Fithian, the "convert or die" president of NATO, agrees with Cameron, Kusama, and the other filmmakers, telling McClintock that "the directors and producers we've talked to are passionate filmmakers and very informed business people who care deeply about their art and craft. Whether they are the makers of blockbusters or independent films, Oscar winners or just starting their careers, all have expressed extreme concern over announced plans to shorten the distribution window, and stated their desire that films can be seen in the venues they were made to be seen in: movie theaters." So the battle lines have been drawn. As the open letter states:

> Make no mistake: History has shown that price points cannot be maintained in the home video window. What sells for $30-a-viewing today could be blown out for $9.99 [or less] within a few years. If wiser heads do not prevail, the cannibalization of theatrical revenue in favor of a faulty, premature home video window could lead to the loss of hundreds of millions of dollars in annual revenue. Some theaters will close. The competition for those screens that remain will become that much more intense, foreclosing all but the most commercial movies from theatrical release. Specialty films whose success depends on platform releases that slowly build in awareness would be severely threatened under this new model. Careers that are built on the risks that can be taken with lower budget films may never have the chance to blossom under this cut-throat new model.

But that "cut-throat model" is how the current theatrical landscape is shaping up, and as McClintock notes in her article, the open letter had "little impact."

With viral marketing campaigns driving the success of such diverse films as *Snakes on a Plane* (2006), *Inception* (2010), *Paranormal Activity* (2009), *Cloverfield* (2008), *The Dark Knight* (2008), *Toy Story 3* (2010), and many others (see McKee for more on this phenomenon), why should viewers stray from their computers to see the actual films themselves? The only thing that supports the theatrical experience is exclusivity; if theaters lose their lock on new releases, they will never be able to recover. Why go out when you can stay home? Why sit with a bunch of strangers in a dark, often dank auditorium when you can relax on your couch and watch the latest releases, even the most mainstream ones? The ads, trailers, teasers, promos, and cast interviews have all gone to the web for streaming download; even high-resolution publicity stills, which used to be doled out to select publications on a case-by-case basis, are now downloadable at the touch of a button. Can anyone really doubt that this shift will eventually (or, more accurately, in the next ten years) happen?

Indeed, attendance at U.S. movie theaters is relatively flat. Overseas, movie theaters are still doing a booming business simply because Hollywood films are so tightly protected by copyright encryption in their new DCPs. As Brent Lang notes, "It's no secret that foreign markets are the one bright spot of growth in the movie business. Last year, according to the Motion Picture Assn. of America, revenue was flat in the U.S. and Canada at around $10.6 billion while the global market reached a record $31.8 billion, driven by a record $11.1 billion overall international grosses. The foreign box office accounted for a record-high 67 percent of all revenue [in 2010]. And [in 2011 and 2012] there's every indication that percentage will go up."

So with theaters in the United States rapidly converting to digital projection in a desperate attempt to hold on to their ever-dwindling market share, it's becoming increasingly clear that even the lure of spectacle, Real 3-D, and IMAX presentation can bring in patrons for only so long. What keeps the theaters open is the exclusivity factor and the studios' reluctance—which seems to be crumbling—to embrace VOD or streaming as the launch platform for new releases. Right now, streaming is still limited to cinema's past, to existing libraries that are being newly exploited for maximum profit potential.

In a way, it's like the 1950s all over again, when Hollywood reacted to the perceived threat of television by rolling out Natural Vision 3-D, CinemaScope, Cinerama, stereophonic sound, and other processes designed to amplify the inherent spectacle of the cinema. Today, studios are ramping up so-called 4-D movies that feature vibrating seats, scents emanating from ventilation shafts inside the theater, swirling fog or other atmospheric effects, and even actors dressed up in costumes to interact with the audience. William Castle, king of the 1950s exploitation gimmicks such as Emergo, in which "the ghost emerges right off the screen" (actually an oversized plastic skeleton drifting through the theater on an overhead wire), and Percepto (vibrating seats that caused audience members to "tingle" during Castle's 1959 film *The Tingler*), would feel right at home with 4-D cinema, which, in addition to all the aforementioned elements, incorporates 3-D projection as part of the package. One of the creators of 4-D cinema, Theodore Kim, optimistically stated that "theaters need to find new ways to bring people back to the multiplex and away from their couches, and this is one way of doing that." Thus, in Ridley Scott's *Prometheus* (2012), as a "giant spaceship crashes into [a] mysterious planet, the seats inside the movie theater heave back and forth and rumble like an earthquake. 'Back ticklers' in the seats thump as an astronaut dodges fireballs and rolls on the ground. A strobe light flashes and huge fans expel gusts of air reeking of smoke and gunpowder" (Verrier). But this sort of thing can't last, just as Castle's reign as exploitation king of the cinema didn't last. They're gimmicks, pure and simple. And people will eventually tire of them, or even avoid them, once the novelty wears off.

In a streaming world, it would seem that for all their splendor, movie palaces have become an anachronism. You can get the whole film online, legally or illegally, for a fraction of the cost of seeing it in a conventional theater. And in a medium driven above all by technological advances and audience demand, the real question is how much longer theaters can continue to survive. Once the studios see a clear and profitable way to cut theaters out of the exhibition equation, just as they did with 35mm film and its attendant technologies, they'll do so in a heartbeat. Perhaps conventional theaters will then become more like museums, where cineastes can contemplate the visions of the past, while mainstream audiences stream the latest blockbuster at home, perfectly content with—perhaps even preferring—the new order of things.

3

Content Wars

Streaming accelerates everything. It creates a voracious appetite for new content. How do you think those 100+ YouTube channels get through the day, providing new programming for their millions of viewers? A look at the promo video on the *YOMYO-MF* (You Offend Me, You Offend My Family) channel demonstrates that it runs like a factory; it's designed to spew out masses of content—good, bad, or indifferent—simply to keep the channel up and running. In many ways, it's like a movie studio in the boom years of the 1940s: some great films were made, but the simple law of supply and demand mandated that many more "program" pictures were produced. Before the collapse of vertical integration in the late 1940s, when many of the major movie studios owned their own chains of theaters, it was essential that a string of cheap films existed to fill theater seats. The same rule applies today.

Content is key to any new enterprise, and although classic films and television programming provide entertainment for many viewers, the vast majority of streaming consumers want new content. We've already seen what happens on the web— mountains of new video, text, and music are added every minute of every hour of every day. This new content comes from a wide variety of sources, but the pressure is particularly acute on established names in popular cinema, pop fiction, and pop music. Books are no exception to this "speed up" in production; indeed, it has made virtual slaves of fiction writers, who used to rely on a more relaxed schedule of production that had been in place for nearly a century. As Julie Bosman notes:

> For years, it was a schedule as predictable as a calendar: novelists who specialized in mysteries, thrillers and romance would write one book a year, output that was considered not only sufficient, but also productive.
>
> But the e-book age has accelerated the metabolism of book publishing. Authors are now pulling the literary equivalent of a double shift, churning out short stories, novellas or even an extra full-length book each year. . . . The push for more material comes as publishers and booksellers are desperately looking for ways to hold onto readers being lured by other forms of entertainment, much of it available nonstop and almost instantaneously. Television shows are rushed online only hours after they are originally broadcast, and some movies are offered on demand at home before they have left theaters. In this environment, publishers say, producing one book a year, and nothing else, is just not enough.

According to Bosman, one author of popular thrillers, Lisa Scottoline, has had to increase her production schedule to Warholian levels, cranking out "2,000 words a day, seven days a week," working nine or ten hours a day. More critically acclaimed writers, such as Jonathan Franzen, "can publish a new novel approximately every decade and still count on plenty of high-profile reviews to promote it," but for mass fiction, like all other forms of popular culture, if you're not continually in the public eye, you run the risk of being forgotten.

At the same time, because the demand for content is so acute, battles over new work erupt on the streaming landscape, as various portals compete for the right to provide the latest film, book, or pop song. I've already discussed the various content providers for movies; for music, iTunes, Amazon, eMusic, Spotify, and CD Baby are some of the primary content providers. And the lines between "provider" and "publisher" are becoming more blurred with each passing day. Amazon, for example, has begun signing up authors of popular novels directly, bypassing publishers entirely. As Richard Curtis, a veteran literary agent, told David Streitfeld, "Everyone's afraid of Amazon. If you're a bookstore, Amazon has been in competition with you for some time. If you're a publisher, one day you wake up and Amazon is competing with you too. And if you're an agent, Amazon may be stealing your lunch because it is offering authors the opportunity to publish directly and cut you out. It's an old strategy: divide and conquer." And so far, it seems to be working. As far as readers are concerned, publishers are not a visible part of the book business; they buy books for their Kindles through Amazon, and that's that. The name of the publisher is immaterial. Indeed, the era of the physical book is over; as of May 2011, Amazon was selling more books in Kindle format than hardcover and paperback versions *combined*. As Chris Gayomali wrote:

Amazon [has] announced that Kindle books are now out-selling both paperback and hardcover books combined. It took a little over two years since the Kindle launched in 2007 to have its e-books outpace hard covers, but it happened in July of 2010 before overtaking paperbacks six months later as well. "Customers are now choosing Kindle books more often than print books," says Amazon founder and CEO Jeff Bezos. "We had high hopes that this would happen eventually, but we never imagined it would happen this quickly—we've been selling print books for 15 years and Kindle books for less than four years."

As Deirdre Donahue notes, "Revenue from the sale of e-books has outstripped revenue from hardcover sales. Using information compiled from 1,189 publishers, the Association of American Publishers reports that for the first quarter of 2012, adult e-books sales were $282.3 million, up 28% from last year. Meanwhile, adult hardcover sales were $229.6 million, up 2.7% from last year. Adult paperback sales were $299.8 million, down 10.5%. During the same period in 2011, hardcover sales revenue was $223.5 million, compared with $220.4 million for e-books." So it seems that the days of the print book are truly numbered, and the shift to streaming reading is not only inevitable but already here.

There are more than a million e-book titles to choose from, most priced at $9.99 or less. But what's interesting is that not *every* book makes it into the Kindle format. Books from small publishers and university presses have a tough time cracking the Kindle market, simply because they aren't designed to be mass media entertainment. Thoughtful, scholarly work isn't what Kindle is after; it wants the latest best seller. Thus the great war has begun in books, just as it has in film—between the texts that are distributed through the Kindle platform and those that aren't and are thus marginalized. This doesn't even begin to consider the numerous out-of-print or public-domain books that might be loaded onto a Kindle, were they available; a very tight list of canonical classics is currently for sale, but as always, the more interesting and adventurous texts are marginalized.

Amazon's Kindle and Barnes and Noble's competing Nook

tablet are clearly the industry leaders and the wave of the future: who has room to store all those hardcover books anymore? Amazon has now become the world's primary purveyor of books, both physical and electronic, placing Barnes and Noble's brick-and-mortar bookstores in jeopardy and forcing the Borders chain out of business. As Amazon has become the front-runner in e-books, Netflix has consolidated its lead on streaming downloads of films and television programs, announcing in January 2012 that its more than 20 million subscribers in forty-five countries worldwide had streamed more than *2 billion* hours of TV shows and movies. That's 666 million hours of on-demand streaming through Netflix every month, or roughly 22 million hours per day (Wauters). These are staggering statistics, signaling that the demand for streaming content can only expand from here.

We've already examined how old television shows can find a second life on the web. But in the case of the cult show *Community*, a half-hour sitcom currently on NBC, a deal was struck with Hulu to stream old episodes on its website and simultaneously underwrite a fourth season of thirteen new episodes in cooperation with NBC. As Meg James notes:

Television production studio executives long have been wary of Hulu and other forms of Internet distribution, fearing they would lead to increased piracy and destroy lucrative secondary markets, including syndication and DVD sales. But video streaming services offered by Netflix, Hulu and Amazon.com are becoming an unexpected boon to the TV syndication market. By writing checks to license library content from networks, the Internet services are injecting new revenue into the TV business and breathing new life into middling shows. "The introduction of the subscription video-on-demand platform has broadened the opportunities for exploitation of product in a very positive way for consumers and studios," said Ken Werner, president of Warner Bros. domestic television distribution. "You do not need to accumulate 100 episodes of a series because 40 hours of programming is a lot, so many of these shows work perfectly well on these new services."

One hundred episodes of any series was the benchmark of a successful foray into syndication. As noted earlier, the producers of Charlie Sheen's new sitcom *Anger Management* hope to stockpile 100 episodes in a mere two years by means of a brutal production schedule that involves shooting a new episode every week, which is unheard of for conventional sitcoms. But with webisodes of streaming-only soap operas and sitcoms clocking in at ten to twelve minutes, a show like *Community*, which is only about twenty minutes long after stripping out the commercials, is a perfect fit.

Not every foray into streaming works out, of course. When ABC canceled the venerable daytime soap operas *All My Children* and *One Life to Live* after four decades, the web production company Prospect Park acquired the rights to both series and hoped to revive them on the web. But after unsuccessful attempts to obtain concessions from the unions to bring stripped-down versions to the web and failed talks with both Hulu and Google, Prospect Park threw in the towel. The main problem: advertisers on the soaps typically appeal to an older audience, and they remained unconvinced that those viewers would follow the series to the web.

But there are numerous "soap operas," or serialized dramas, on the web: *Gotham* (set in New York City, naturally), an upscale drama centering on the lives of young professionals; *The Lake*, produced in concert with actor-producer Jason Priestly and webcast on WB.com; as well as *In the Moment, Wed-Locked, Buppies* on Bet.com, and *Foreign Body*, all of which are pioneering efforts that may or may not catch on. You can easily look up episodes and see for yourself. In keeping with web viewers' limited attention spans, most episodes of these series run less than ten minutes, and they suffer from being "stand-alone" projects rather than being part of, say, the YouTube "network" of channels, where viewers are almost compelled to surf through the variety of offerings on display, all conveniently located on one site—like flipping through the channels on a conventional television set.

But since content is king, there are continual skirmishes for the rights to premium properties, such as when Starz and Netflix battled over the price of content that Starz provided— which was really Starz's attempt to hang on to its cable customers rather than seeing them migrate to Netflix (Fritz, Flint, and Chmielewski). At the same time, iconic Hollywood films based on public-domain literary sources are also coming under scrutiny. As Eriq Gardner notes, for example, "In 1900, L. Frank Baum wrote the famous children's novel, *The Wonderful Wizard of Oz*. It's easy to assume that since the book was published more than 110 years ago, the characters of Dorothy, the Scarecrow, the Cowardly Lion and the rest are safely in the public domain." And while that's true, it goes only so far. The characters as *written* by Baum may be free for public use by anyone, but the characters as depicted in the 1939 movie are not. Warner Bros., which now owns the rights to that movie version of *The Wizard of Oz*, won a decision to that effect in early July 2011. There are, right now, at least four or five proposed Wizard of Oz projects in various stages of production in Hollywood, but all of them will have to steer clear of any visual depictions of Baum's characters as popularized by the film.

This is not exactly new territory. In 1957 Hammer Films in England decided to create a horror film, *The Curse of Frankenstein*, based solely on Mary Shelley's 1818 novel. Universal Pictures, which had produced the 1931 version of *Frankenstein* starring Boris Karloff, vociferously objected, stating that the characters of Frankenstein and his monster were its exclusive property. Hammer disagreed, noting that the novel was clearly in the public domain, and it scrupulously examined every Frankenstein film made by Universal to ensure that no thematic or visual elements from the Universal series appeared in its production. By sticking closely to Shelley's source material and religiously eschewing any aspect of Universal's films, Hammer came up with an entirely new depiction of the monster (as played by Christopher Lee) and of Baron Frankenstein (Peter Cushing) and succeeded in creating a completely original concept of the Frankenstein story. In fact, after the film opened on a worldwide basis through Warner Bros. and did extremely well at the box office, Universal decided to license its entire catalog of cinema monsters to Hammer for remakes, in return for world distribution rights and a healthy chunk of the profits. The first was Hammer's 1958 production of *Dracula*, released as *Horror of Dracula* in the United States.

So, public-domain characters are still fair game for filmmakers, as long as they stick to the source material and avoid using any elements of previous film adaptations. Thus, a flood tide of new versions of public-domain works is one easy way to supply "presold" product to the viewing public.

And the "content wars" aren't confined to programming possibilities or public-domain characters or even the web's need for unceasing production. They are also about the "real estate" of theatrical presentation in an age of streaming imagery. When 20th Century Fox, Sony, Warner Bros., and Universal all threatened to push their films onto VOD platforms much sooner after their runs in theaters, one theater chain, Regal Cinemas, responded by cutting back on the screening of trailers for the studios' upcoming films. Regal reasoned that if the studios embraced VOD so enthusiastically, this would undermine theatrical distribution. Who would go to see a film in a theater if, within sixty days, you could see that same film as streaming video?

For some of the smaller companies, like IFC, the VOD strategy makes sense. IFC specializes in foreign imports and relatively low-budget films that open in only a few major cities. IFC long ago established a "same-day" policy, offering its films as VOD downloads on the same day they open theatrically (sometimes in as few as *two* theaters). By charging $9.99 for the same-day release of a film that will never gain national commercial traction, IFC established a useful model for independent distributors: open in a few theaters for the prestige value, such as in New York and Los Angeles, so that the film is reviewed (the *New York Times* is especially good in this regard, offering cogent and informed critiques of nearly every film that opens in Manhattan, no matter how modestly budgeted or poorly publicized), and then offer the film nationwide on cable systems, such as Time Warner Cable, so that cinephiles who otherwise would never have a chance to see the film can view it immediately.

But this system isn't effective for major releases: a $9.99 price point simply won't generate the financial returns necessary to drive films like *The Dark Knight Rises* to box-office dominance and, more importantly, recover production and distribution costs. Even if 35mm prints are obsolete and the studios save a fortune in shipping and print costs, there's still the all-important publicity blitz to consider, as well as the creation of a multiplicity of trailers, promos, and "making of" shorts, all of which are part of the promotional process.

Theatrical real estate thus remains a prickly problem: the major studios need it for their tent-pole films (those that dominate at the box office for several months, grossing enough money to offset the financial losses of weaker films), but they hate the fact that they *still* have to go through a middleman. The studios would like to control the entire process from start to finish, as they did in the days when Adolph Zukor's Paramount Publix Corporation controlled the majority of the nation's theaters, as well as the world's largest distribution network. Then the government stepped in and ordered Paramount, as well as the other studios, to choose between operating as production units and operating as exhibitors. Paramount and the other studios that owned their own theaters fought what became known as the Consent Decree for more than a decade, but it was always a foregone conclusion that they would stay in the production side of the business. That way, they could dictate their terms to the theaters, which is what they did then and what they're still doing now. But the odd symbiosis of the studio-theater relationship remains strained; each needs the other, yet each fervently wishes this was not the case.

In late 2011 I interviewed rising director Dale Resteghini on this topic, and his response crystallizes the current state of affairs regarding theatrical distribution—a sort of Mexican standoff in which no one is really the winner. As Resteghini observed:

Ultimately, theatrical is still going to be where the studios get the big, big dollars, because every studio needs that big opening weekend to push the film over the top, which is what they build their marketing plan around. You need fan buzz build-up before the opening, or otherwise you lose out on kids texting each other: "I just saw this movie, and here's what I thought" right while they're watching the film! They want the moviegoing experience, sitting in the dark, escaping the real world for a couple of hours. I believe the "real estate" of theatrical presentations, in both the physical and theoretical sense, of being able to show a film with a big build up over a weekend, opening on Friday, is the only way to make sure you get as much money out of the film as possible. The studios need to hold on to that, because otherwise it's going to become like music, where there are so many studios releasing films that new films are going to get lost. So as much as these other new technologies are good for ancillary distribution, for mainstream, big budget films, theatrical is the only way to go. (Dixon, "Sleepers," 11)

And then there's always Redbox, a primary distributor of content for those who can't afford or don't want to be beholden to cable television or the web. But it has endured its own content wars. By now, you've probably seen one of the ubiquitous Redbox DVD vending machines that populate the various malls around the nation, and perhaps you've even had occasion to use one. Most Redbox DVDs are "top forty" hits, so to speak, aimed squarely at the lowest-common-denominator viewer and incessantly chatted up in gossip columns and other media outlets that hype the latest releases. These machines are remarkably popular, and they should be. As Brooks Barnes noted in 2009, they are the result of a once-failed entrepreneur retooling his marketing strategy to reach a wider audience:

In 1982, just as the VHS tape was taking off, a *Star Wars* buff named Mitch Lowe had a radical idea. What about building a vending machine that could rent movies? He called his invention "Video Droid." It failed. People were not yet comfortable using credit cards for casual transactions, the tapes broke easily and the technology involved with manipulating their bulk proved too expensive. But Lowe did not give up, and his moment seems to have finally come. Lowe, 56, is now the president of Redbox, a fast-growing company in Illinois that rents movies for $1 a day via kiosks. By December [2009], there will be 22,000 Redbox machines in spots like supermarkets, Wal-Mart stores and fast-food restaurants.

It's also interesting to note that Lowe was a cofounder of Netflix and spent some time at McDonald's as well, working in marketing (Tribbey). Indeed, Lowe is currently working on a deal to put Redbox kiosks in McDonald's restaurants, which seems perfect: fast food and fast movies under the same roof.

Redbox has been, by any measure, an extremely successful marketing strategy. By 2007, Redbox kiosks were the fifth largest rental operation in the United States (Mui). By 2008, Redbox kiosks had rented out more than 100 million DVDs ("Redbox Surpasses"); in 2009 that figure ballooned to more than 400 million rentals (Tribbey). As of May 6, 2010, the company had 25,000 kiosks in place, and the number is still growing. These kiosks average roughly fifty transactions a day, and DVDs are rented (and returned) approximately fifteen times before being consigned to the scrap heap (Mui). Rentals are cheap at $1 a day. Most customers return the DVDs to the kiosks the next day, but for those who don't, their credit cards are charged for a certain number of days and then the consumer "owns" the disc, minus the box and any artwork or liner notes. In addition, customers can now reserve DVDs online from their favorite Redbox location and pick them up later, although most customers are walk-up users who simply troll through the menu to see if anything interests them (Lackey).

There were, of course, bumps along the way to Redbox's dominance in the self-serve DVD market. A flurry of lawsuits ensued as the movie studios attempted to wrest some of the rental profits from the company and Redbox sought to preserve its autonomy as a rental facility. As Chris Tribbey reported in 2009,

Redbox "reached a five-year, $460 million agreement with Sony Pictures Home Entertainment. When Universal Studios Home Entertainment tried to establish revenue-sharing terms for its DVDs, Redbox took the studio to court." All this legal maneuvering still has to shake out, but it seems clear that, for the moment, Redbox is here to stay.

The most curious thing about Redbox is that it exists at all: why are people still renting DVDs when they can so easily stream films or download them to their computers? Blockbuster has been plowed under, for all intents and purposes, and Redbox is essentially a "brick-and-mortar" operation, even if, like Coke machines, it occupies existing store space, lobbies, and parking lots by mutual agreement. Further, it deals in DVDs, which are rapidly becoming an obsolete format. How does Redbox do it? Mitch Lowe offered these thoughts about the company's success in an increasingly digital landscape in a 2009 interview: "Redbox growth can be attributed to a combination of factors. On the consumer side, Redbox offers the convenience of providing consumers with access to DVD rentals where they shop, value, and an easy, simplistic rental experience. On the business side, our retail partners benefit from the value in revenue share and increased consumer visits generated by Redbox." And the $1 price is clearly a factor; as Lowe notes, "Redbox has helped revitalize the DVD industry by creating incremental rentals among viewers who previously were priced out of renting and buying DVDs." As for the digital, streaming future, Lowe says, "There is no question that electronic delivery of content is gaining in popularity . . . [but] Redbox is well positioned to succeed in both the digital media and physical media space. Our future strategy encompasses both 'clicks' and 'bricks.' We have what no other digital distribution company can claim: a physical presence through our retail partners at over 17,000 locations and growing. And, we are uniquely positioned to host conversations with our customers throughout their consumption of entertainment media, in whichever format they prefer" (Tribbey).

That's pretty slick corporate speak, but what do you expect? Redbox is essentially a down-market operation, aimed at the consumer whose only knowledge of film is what she or he reads in the entertainment section of *USA Today* or *Cosmopolitan*. A recent informal check of the inventory of titles at a local Redbox confirmed that the films are major studio releases of one kind or another, mostly (of course) genre films, and genuine theatrical hits are mixed in with films hoping to break even with DVD sales. The other factor to consider is that many people who use Redbox as a home entertainment option don't have or can't afford cable and perhaps don't use a computer very much. They have DVD decks and may be cautiously moving into Blu-ray, they play video games—an area Redbox is clearly interested in—and they watch television via satellite or even an antenna. Not everyone is wired in, even in 2012, even though more and more consumers become part of the digital world every day.

Battles over this territory remain to be fought, and they will be intense and perhaps ongoing. Everything is in a constant state of flux, and new delivery systems—unimagined at this moment, except by a few—may well render the formats considered here obsolete. Redbox may have a considerable hold on the mainstream DVD rental market for the present, but it remains to be seen how the company will perform in the streaming future. When everyone goes online for everything, even to buy groceries (it's happening), there may be no reason to leave the house, and Redbox's numerous mall locations may no longer be so enticing. For the moment, however, Redbox has established a comfortable niche and is continuing to expand, even as digital clouds gather on the horizon.

And then we come to the issue of video games, which have seemingly taken over the web and attract legions of users. Along with Facebook, online streaming video games seem to have hypnotized hundreds of millions of people on a continual, daily basis. In the 1960s an oft-heard mantra was "Be here now"—center down, appreciate your current existence, be sensitive to your real surroundings. Now, with *Farmville*, *World of Warcraft*, and other online role-playing games too numerous to mention, to say nothing of online gambling, porn, and poker sites, the motto might as well be "Be somewhere else—anywhere but here." Social networks are one narcotic escape, but massively multiplayer online role-playing games (MMORPGs) are equally addictive to much of the population. I don't think this is a good thing.

As Wikipedia defines them, succinctly but all too accurately, "massively multiplayer online role-playing games are large multiuser games that take place in *perpetual* [my emphasis] online worlds with hundreds or thousands of other players. MMORPGs can also include computer role-playing games in which each player controls an avatar that interacts with other players, completes tasks to gain experience and acquires items." As with Facebook, the idea is to always be online, living your life entirely in a virtual world. The attraction, presumably, is that your own life is so empty that anyplace is better than where you really are. Also, because you lack the creativity to dream up imaginary worlds of your own, digital or otherwise, you sign up to participate in someone else's fictive construct for unlimited amounts of time.

Of course, one could argue that a similar phenomenon occurs in a movie theater: you sit in a darkened auditorium with hundreds of strangers and submit to whatever is presented on the screen. The same is true when you watch conventional television. But these are one-way experiences; unless you're a theorist who is deconstructing the images presented to you, or a lay viewer entirely caught up in the narrative, you can always switch the television off or leave the theater. And television is a passive medium in any event, almost designed to serve as background information while you get on with your life in the real world.

In interactive games and social media websites, you have the illusion of control, in that you can manipulate avatars to do your bidding (as long as those avatars remain within the confines of the game) and add or delete "friends" at will. But outside the game or the social media site, avatars and friends cease to exist, and so does the world they inhabit. What happens to the game player who has spent an enormous amount of time online? Does he or she cease to exist as well? And what about your virtual "friends"? Are they your friends when you're not online? The other factor is quite obvious: you are the "star" of your own Facebook profile, so the world becomes very small and insular, as if you're someone special. But you're not. You're just one of the 900 million people on Facebook. Facebook, of course, wants to personalize the experience and make you feel like genuine communication is going on. But all that's really happening is data mining and a series of tenuous online "relationships" that can evaporate with a few keystrokes.

And the online world can be awfully seductive. In 1956 Charles Eric Maine (born David McIlwain in 1921) published a superb, often overlooked science fiction novel titled *Escapement*, which posits a bleak virtual future. In Maine's novel, tech mogul Paul Zakon, head of the 3-D Cinesphere Organization, builds a worldwide network of "Dream Palaces," where millions of "dreamers" lie immobile in isolation chambers, maintained in a semicomatose state through a combination of intravenous drugs and liquid protein. These dreamers are hooked up to electrodes that feed them scenarios more real than life, allowing them to experience a simulated existence of power, wealth, and sexual abandon. These men and women spend most of their lives in a fantasy world, from which they emerge only when their money runs out and they are taken out of the system. Then, like the addicts they are, the erstwhile dreamers work desperately at whatever menial jobs they can find until they manage to scrape together enough cash for another six months or so in one of the Dream Palaces, and the process starts all over again. As Maine prophetically writes (more than half a century ago), describing the rise of Zakon's Dream Palaces:

> At first the thing had been a novelty, an expensive novelty, demonstrated in a handful of specially adapted theaters in the major cities of the States. But the novelty had also been an enormous success. The Cinesphere studios converted their sound stages into psycho-recording sets, and ambitious productions were recorded on miles of brown plastic tape. Lavish, spectacular and sensational productions, loaded with romance and glamour and an aphrodisiac innuendo of sex. . . . In the space of four years psycho theaters—later to be called Dream Palaces—were installed in their thousands throughout North America. . . . Dreamplays were produced that ran continuously for days, and then weeks, and finally, years. (182–84)

Zakon's unwilling associate is Dr. Philip Maxwell, whose research led to the creation of the Dream Palaces. Maxwell becomes increasingly uneasy about the growth of Zakon's empire and starts to move against his employer, but he finds that Zakon's hold on both the populace and the law is too tight. People *want* what the novel terms an "unlife"; otherwise, why would it be so popular? Eventually, a quarter of the world's population is sequestered in isolation tanks, and as they increase the length of their dreams, they default on mortgage payments and other responsibilities. The Cinesphere corporation acquires their property and cash savings, exponentially increasing Zakon's empire with each passing day. As he tours one of the facilities with Zakon, Maxwell stops to examine the isolation tank of one Paula Mullen, age twenty-seven. She has signed up for a dream entitled "woman of the world"—eight years of uninterrupted synthetic fantasy in which she imagines herself alive, awake, and the center of worldwide media attention. In reality, of course, she is an immobile, corpse-like husk in an oversized filing cabinet, but Zakon sees nothing wrong with this. As he tells Maxwell:

> There's nothing anti-social about unlife, Maxwell. In fact, it acts as a scavenger of society, and removes the more anti-social types from active circulation. Take this Miss Mullen . . . and try to imagine her as a useful member of society. She chose to escape from society for eight years. That proves she was one of the many millions of maladjusted people living out their lives in dull unending routine. The kind of people who find no creative pleasure in work. Who seek their fun in furtive sex relations and objective entertainment. She's better off here. She's happier than she ever knew and she's no longer a burden to society. (207)

By the end of *Escapement*, Maxwell's revolt is successful. He brutally murders Zakon with a jack handle and records Zakon's experience of death on one of Cinesphere's dream machines. Then, intending to awaken the millions of dreamers, Maxwell forces his way into the central control room of Cinesphere and orders the technicians to play the "psycho tape" of Zakon's murder throughout the entire system. But the tape of Zakon's death is *too* realistic; nearly 100 million dreamers ensconced in Cinesphere's isolation tanks die along with him, unable to endure the horror of being beaten to death. Maxwell is put on trial, and the prosecution sums up the charges against him:

> Ninety-eight million, four hundred and thirty-two thousand eight hundred and twelve people, men and women, old and young, died suddenly one evening nine months ago. They were voluntary dreamers seeking relaxation in the many Dream Palaces of the Zakon organization. This man, Philip Maxwell, coldly and brutally murdered Paul Zakon, and made a psycho-recording of his death agonies which he then played back over the world unlife network. The result you all know. Psycho dreams are realistic. Psycho nightmares are equally realistic. And psycho death is indistinguishable from ordinary death. (221)

The book ends with Maxwell sitting alone in his prison cell, awaiting the final verdict.

While the conclusion of Maine's novel is undeniably melodramatic, comparisons between the operations of the fictional Cinesphere corporation and the real-life virtual worlds offered on the web are hard to resist. Hundreds of millions of people subscribe to these virtual worlds and spend countless hours in them, and they pay real (as opposed to virtual) money to "exist" in a more attractive alternative universe. Although digital technology has far surpassed the mechanics of Cinesphere's fictitious operations, the fact remains that for many, online virtual life has become an addiction; it has become more "real" than the physical existence they so desperately hope to escape. *World of Warcraft*, for example, currently boasts 10 million users and counting, and its advertising motto is "10 million people can't be wrong." *Farmville* has some 33 million users, with more subscribing every day. Nine hundred million people are on Facebook, which encourages its users to "be somewhere else" rather than living freely conscious lives unplugged from their computers. Can 900 million people be wrong? Indeed, they can.

Legendary filmmaker Luis Buñuel recognized this (with typical prescience) when he decried in 1980, just three years before his death, the ceaseless profusion of meaningless images that confronts us at every turn. And this was long before the digital revolution took hold. As he noted in his essay "Pessimism": "The glut of information has . . . brought about a serious deterioration in human consciousness today. If the pope dies, if a chief of state is assassinated, television is there. What good does it do one to be present everywhere? Today man can never be alone with himself, as he could in the Middle Ages. The result of all this is that anguish is absolute and confusion total" (263). If you're listening to an iPod, you're not here. If you're on Facebook, you're not here. If you're on *Farmville*, you're not here. If you're playing an interactive video game or an MMORPG, you're not here. If you're checking your e-mail or your Blackberry or you're tweeting, you're not here. If you're not here in the here and now, then you're not here. Virtual realities don't exist; virtual unrealities do. And when you check into them, you become unreal as well.

No one would want to go back to the analog era and give up the many conveniences (especially in the world of moving image studies) the digital world can offer. And yes, you could easily say that reading a book or listening to a CD or watching a movie is similar to being online and that for the duration of those experiences, you aren't here, either. But those experiences have limits, whereas the goal of Facebook and all its allies is to get you online and keep you online—forever, if possible, as in "keep me logged in." And when that happens, as far as I am concerned, you cease to exist in the real world. Or as Scott Adams put it recently, "Someday all of our technology will learn to emotionally manipulate us. Your smartphone is already doing it. Your desktop computer has been doing it for years. As your possessions learn to fill your emotional void, your need for the comfort of other humans will continue to decrease. Eventually we'll be a society of sociopaths. I'm already halfway there."

Yet, despite all these technological inroads—or perhaps I should say invasions—into our lives, consider these much more cheerful statistics:

- Seventy-five percent of conversations in the United States (and even more in other countries) still happen face-to-face; less than 10 percent take place through the Internet.

- Face-to-face conversations tend to be more positive and more likely to be perceived as credible, in comparison to online ones.

- In the sphere of products and services, conversations are significantly impacted by what we see and hear in "traditional" media, including television, radio, and print publishing. (Keller and Fay)

In a world where you can create an imaginary "Cloud Girlfriend" (see Considine for more on this) to keep you company if you're utterly detached from the real world, these markers of the value of actual existence are both reassuring and necessary reminders that while we live in a landscape that is becoming more and more mediated by technology, it need not overwhelm us. As Sherry Turkle, a psychologist and professor at MIT, recently noted:

Texting and e-mail and posting let us present the self we want to be. This means we can edit. And if we wish to, we can delete. Or retouch: the voice, the flesh, the face, the

body. Not too much, not too little—just right. Human relationships are rich; they're messy and demanding. We have
learned the habit of cleaning them up with technology. And
the move from conversation to connection is part of this.
But it's a process in which we shortchange ourselves. Worse,
it seems that over time we stop caring, we forget that there
is a difference. We are tempted to think that our little "sips"
of online connection add up to a big gulp of real conversation. But they don't. E-mail, Twitter, Facebook, all of these
have their places—in politics, commerce, romance and
friendship. But no matter how valuable, they do not substi-
tute for conversation. (SR6)

We can think of our own communications, therefore, as part of
the "content wars," with our edited selves as scripted as any television show or PowerPoint presentation, becoming part of the
media landscape. Or we can think of ourselves as separate entities, distinct from the technological world, and step back from it
just a bit. A life online is no life at all. It is merely an information
stream—a content stream—as manipulated as any media offering, no matter what the format. We are real, corporeal, human,
and mortal. We drive the machines—not the other way around.

In May 2012 the Internet Week conference in New York co-
incided with the television "upfront" presentations, where networks unveil their new prime-time offerings. As Brian Stelter
and Stuart Elliott noted, the web and conventional television
seemed more connected than ever:

At the upfronts, network TV executives spoke the languages of social media and web science to a greater degree than
ever. ABC spoke about designing great "user experiences"
just as a web designer would; both Fox and CBS showed off

how many online fans they had on sites like Facebook and Twitter. Fox says it has 230 million; CBS, 167 million . . . [B]ut the real take-away [was Les Moonves's] Carnegie Hall presentation on Wednesday [which introduced] the phrase "first screen first." As the image of a flat-screen television was projected above him, Moonves [the CEO of CBS] said, "We love social media enhancing our product, but most importantly, with all of this, everyone is still talking about the first screen [i.e., the conventional television screen] first."

And yet Anne Sweeney, president of the Disney/ABC Television Group, started ABC's upfront by saying, "Our audience is tuning in, logging on, streaming, downloading—devouring our content any way they can get it" (as quoted in Stelter and Elliott), persuasively demonstrating just how interconnected television and the web have inevitably become.

As the demand for content inexorably expands, the acquisitions keep coming, and companies swallow one another in a desperate race to remain vibrant competitors in the marketplace. Blockbuster, the pioneering video chain that dealt primarily in DVDs, is having a hard time escaping its past. It was purchased by Dish Network—a satellite alternative to cable—for a mere $320 million in April 2011 (Fleming), a fraction of what the company was worth at the height of the brick-and-mortar era. The acquisition is a troubled one, to be sure. Many Blockbuster stores throughout the United States have closed as a result of the swift move to streaming video, although Blockbuster now offers a streaming service of its own. Yet in the public's mind, Blockbuster is still the place where one lines up to rent DVDs. And in a Redbox world, Blockbuster is finding it difficult to adjust to the new consumer marketplace.

Meanwhile, Netflix has been aggressively acquiring content

for its library from the major studios. It now streams episodes of *Glee, Sons of Anarchy, Ally McBeal,* and *The Wonder Years* to its millions of subscribers, bypassing cable outlets altogether. In the area of popular music, Apple's iTunes division signed pacts with Sony, Warner Music Group, and EMI to stream their catalogs of songs to users of iPhones, iPads, and conventional computers, and it purchased the domain name iCloud for a reported $4.5 million to help "brand" the service (Kolakowski). By storing their music in Apple's "cloud" computing system, users free up space on their own hard drives for other content; Amazon's Cloud Player offers exactly the same service to its customers. This is, of course, a double-edged sword for record companies: artists no longer have any "label identity"; they're all available on Amazon or iTunes. Thus, the record labels have become less of an entity unto themselves; they are merely content providers for streaming distribution servers.

Young actors and entrepreneurs dominate the streaming world, as one might expect, with its focus on what's new, hot, and trendy—what's "in the moment." Actor Ashton Kutcher has become a ubiquitous presence in the streaming community, investing in the social networking service FourSquare and, not coincidentally, cross-plugging it with product placements on his hit television show *Two and a Half Men* (in the form of FourSquare decals on his character's laptop). He has also provided capital for Path, a photo-sharing application, and Flipboard, a "news reading" application for the iPad. In 2009 Kutcher invested a sizable chunk of cash in Skype, then valued at $2.75 billion; in 2011 Microsoft bought Skype for a figure north of $9 billion, so Kutcher clearly has a knack for spotting hot tech start-ups.

As Kutcher puts it, "I look for companies that solve problems in intelligent and friction-free ways and break boundaries," which seems a modest self-assessment of his business acumen

(Wortham). Kutcher, of course, has a large following on Twitter and Facebook, which he uses to publicize his activity (although, after some embarrassing comments about the firing of football coach Joe Paterno, he turned over the day-to-day management of his Twitter account to his publicists). Yet through it all, Kutcher manages to maintain a calm yet high-profile presence in the streaming world, appearing at digital media conferences such as Tech Crunch Disrupt and confining his public remarks on his investment practices to homespun homilies like this: "I have a bunch of interesting and really smart people that I sit with and talk to quite frequently because of the investments I've made, and between their networks and mine, I get to see things really early" (Wortham). Indeed he does, and he's making it pay off.

At the same time, as we move to the boundaries of what's possible in the streaming universe, it isn't all about downloading or uploading music, movie, or text files. Google and Facebook have both been dabbling in facial recognition technology, in which the photos of customers, or "members" of their online communities, are scanned for instant identification, leading to serious privacy concerns in the view of many experts. Google stated publicly that it has decided not to go forward with this technology, but as Sarah Jacobsson Purewal commented, Facebook is actually better poised to create and release an accurate facial recognition search system—and nobody from Facebook has stepped forward to say that it won't do so.

Indeed, Facebook is interested in facial recognition software because it wants to track and tag everyone who logs on and uncover more information about them. Your face is the most valuable data you own, because in the future, when passwords become obsolete, your face will serve as the password to log on to all your electronic devices. And in the near future, if your image is pirated, it will be the equivalent of stealing your Social Security number: it could potentially unlock your whole life for anyone who wants to gain access to your world. As Jared Newman noted in June 2012, Facebook "acquired Face.com, a facial recognition start-up, in a likely attempt to make photo-tagging easier on the social network. . . . Face.com's technology can identify Facebook users' faces in photos or live video. The company already offers a Facebook app called Photo Tagger, which can identify faces and suggest photo tags, and also offers a standalone iPhone app called Klik that can identify friends in real time and adapt image filters to people's faces."

Thus, Facebook is now collecting the most valuable information of all: the essence of your being. It will possess your likeness and all the data it unlocks, and it could sell your image to anyone who might find it profitable. And with Facebook's 900 million–plus members, it's off to a good start. If you add in all the photos of family and friends that have been posted by all these members, Facebook probably has at least 2 billion or 3 billion faces on file for facial recognition software to sort through at lightning speed. Soon the population of the entire planet may be visually cross-indexed and the results sold to the highest bidder. As Ed Oswald explains:

If you're worried about certain pictures of yourself on Facebook, now's the time to delete them. Security firm Sophos issued a warning on Tuesday saying the social networking site had enabled facial recognition technology on accounts without informing users of the change. . . . Facebook used facial recognition technology to prompt your friends to tag you—which means photos of you are *more likely* to be tagged. After all, people are more likely to tag someone if Facebook pops up with a notification and suggests who that person is. . . . The new tagging isn't much different from the way things used to be—friends could always tag photos of you, and you would have to untag yourself manually. The difference is that the process is now semi-automated, and some may find this an affront to their privacy.

Purewal agrees and notes that with more than 90 billion photos (as of June 2011—there are certainly many more now) in its database:

Facebook's facial recognition technology is downright creepy. Opting out of the service doesn't mean Facebook will stop trying to recognize your face—it just means that Facebook will stop suggesting that other people tag you. . . . Facial recognition technology will ultimately culminate in the ability to search for people using just a picture. And that will be the end of privacy as we know it—imagine, a world in which someone can simply take a photo of you on the street, in a crowd, or with a telephoto lens, and discover *everything about you* on the internet. ("Why Facial Recognition Is Creepy")

Thus, when considering "streaming content" on the web, it's a mistake to confine one's thinking to conventional files of preexisting material, whether it's music, video, or text. The most valuable commodity in the streaming world is personal information, a vast database of consumer preferences and desires that is jealously guarded by those who compile it, constantly updated, and sold to anyone who wishes to target a specific audience—or the *entire* audience—on the web. Streaming isn't just about downloading or uploading files. Streaming is about the merchandising of one's identity for profit, because without a database of consumer information and a list of subscribers, what would any of the social networking services really be worth? *Nothing.* It's users who give value to Facebook, LinkedIn, FourSquare, and their numerous progeny.

So it's small wonder that managing one's online presence has become big business, because the streaming universe never forgets. Once something goes online, it stays there forever. If it appeared in a newspaper, a magazine, or even a newsletter, someone has probably scanned and archived it on the web, where it is available for immediate downloading. Researchers used to spend days, weeks, and months compiling information from obscure journals and local news sources on a given topic; now, it's all available with the touch of a button. This existing content from other "old-school" media constitutes one of the key databases on the web, which has neatly absorbed all previous forms of publication and information dissemination and made their content proprietary information to be rented or sold to the highest bidder.

In response to this, Google launched a somewhat lukewarm "tool" in 2011 to help web users manage their online reputations, noting that "your online identity is determined not only by what you post, but also by what others post about you—whether a mention in a blog post, a photo tag or a reply to a public status update" (as quoted in Kessler). Google's official announcement of the service adopts a folksy tone: "Nowadays, more and more personal information surfaces on the web. For example, some of your friends might mention your name in a social network or tag you on online photos, or your name could appear in blog posts or articles. Google search is often the first place people look for information that's published about you." But then it offers a less-than-reassuring laundry list of what one can do to combat negative posts on the web. It turns out that it isn't so easy to remove unwanted citations of your name, history, age, place of birth, education, employment, photograph, or other identity markers, *unless you control the content of the site itself,* which of course, in most cases, you don't. Notes Google somewhat guilelessly:

If you find content online—say, your telephone number or an embarrassing photo of you—that you don't want to appear online, first determine whether you or someone else controls the content.

For example, if the photo you want to hide is part of your Picasa account [which almost no one has], you can simply change your photo visibility settings. If, however, the unwanted content resides on a site or page you don't control, you can follow our tips on removing personal information from the web and removing a page from Google's search results.

But this is easier said than done. As Google admits:

It's difficult to keep personal information off the web. Most people belong to social networks, post photographs, write blogs, tweet, or have personal or professional sites. But this doesn't mean that you always want this information to appear in search results. . . . If you'd be uncomfortable with a photo or piece of content being visible to strangers, think carefully about publishing it on public sites. If something compromising has already been published, try to remove it from the site where it's appearing. . . . If you can't get the content removed from the original site, you probably won't be able to completely remove it from Google's search results, either. Instead, you can try to reduce its visibility in the search results by proactively publishing useful, positive information about yourself. (Google Accounts Help)

After all, Google doesn't "own" the web. No one does. What the web represents is a vast repository of original and archival material in all possible formats, waiting to be used by someone, somewhere. Content providers simply want paying customers, and as long as you have a credit card or a PayPal account, you can stream pretty much any information or entertainment material you want, legally or illegally. Indeed, some file-sharing sites that are transparently illegal *still* charge user fees, and they also collect information about all those who frequent the site.

The content wars are thus ubiquitous and continually exploding throughout the web, like sunspots that erupt periodically with volcanic fury and often with little or no warning. Content providers—whether they be film production companies, video producers, webcast creators, music libraries, repositories of printed texts, or websites for online games—continue to proliferate with almost unimaginable speed. Hundreds of new sites pop up every day, whether legal or not, and many of them are "aggregators" of existing content that "repurpose" material created by others, often without consent. Without content, there would be no web culture, so the competition continues to increase on a minute-by-minute basis. There is little anyone can do to control it or manage it; indeed, if one tried to absorb all the material available on the web, it would take multiple lifetimes to do so.

4

The Moving Platform

As the web continues to expand, like the universe, into a maelstrom of almost inconceivable complexity, it's important to remember that there is a life apart from virtual existence, a real world where we possess corporeal existence. We ignore this at our peril, and the risks of video game addiction or Facebook's demands on our time come with an additional warning: too much time online can, in and of itself, be hazardous to our health. This is particularly true of watching television, no matter what the source of the content is. As Crystal Phend discovered:

A couple of hours of daily TV watching can add up to substantial risk of type 2 diabetes, heart disease, and death, according to a new review of past research. Every two hours per day spent in front of the television booted relative risk 20 percent for type 2 diabetes, 15 percent for cardiovascular disease, and 13 percent for all-cause mortality, Anders Grontved of the University of Southern Denmark in Odense and Dr. Frank B. Hu of Harvard School of Public Health found. Hu suggested that adults should limit themselves to no more than two hours per day. . . . TV watching isn't just time taken away from less sedentary activities, but also is often time spent eating or snacking and watching commercials for unhealthy foods, Hu [said, noting that] for [every] 100,000 individuals watching two hours per day over the course of one year, 176 would become diabetic, 28 would die from cardiovascular disease, [and] 104 would die from any cause.

For those watching three or more hours a day of television, the "risk of early death rose even higher" (Phend). With Netflix, Hulu, and other web content providers now streaming movies and television shows through the family TV set—often with commercials, as in the case of Hulu—the risks become substantial. And yet, as Nielsen reports, Netflix users are streaming more television than ever and fewer movies:

> The increase comes in the wake of Netflix not just inking partnerships to re-air programming, but also jumping feet first into production themselves, rolling out original content—like the 8 episode *Lillehammer* starring former Soprano sidekick Steven Van Zandt or the upcoming *Orange Is the New Black*, which is the first project from *Weeds* showrunner Jenji Kohan following the Showtime program's run. . . .
>
> Conversely, subscribers mainly streaming movies dipped from 53 percent in 2011 to 47 percent in 2012. ("Survey: Netflix")

And as all those television sets find new programming via the web, more and more viewers are cutting the cable cord, which is exactly what Cox Cable and Time Warner Cable are afraid of, and it's also why these companies and others encourage customers to bundle all their services into one package. The old sales pitch for a combined telephone, cable, and Internet package stresses economy and convenience, but it also allows cable companies to monitor customers' usage of the web and see exactly how much material they're streaming versus viewing on traditional cable networks. In addition, it's a temporary way to stop people from abandoning cable altogether—making it an "essential" part of your home service package when, in reality, you could easily do without it. As Peter Svensson notes:

> People are canceling cable, or never signing up in the first place, because they're watching cheap Internet video [so] viewers can expect more restrictions on online video, as TV companies and Hollywood studios try to make sure that they get paid for what they produce. . . . Ian Olgeirson at [research company] SNL Kagan . . . expects programmers to keep tightening access to shows and movies online. A few years ago, Olgeirson said, "they threw open the doors," figuring they'd make money from ads accompanying online video besides traditional sources such as the fees they charge cable companies to carry their channels. But if it looks as if online video might endanger revenue from cable, which is still far larger, they'll pull back. "Are they really going to jeopardize that? The answer is no," Olgeirson said.

Viewers aren't the only ones abandoning cable. Larry King, a longtime veteran of CNN, quit his talk show *Larry King Live* after more than twenty years and moved to Ora TV, a project financed by Mexican billionaire Carlos Slim. King now stars in a four-night-a-week series entitled *Larry King Now*, which is just like *Larry King Live* except that it's half an hour long and available only on the web. Nevertheless, thanks to King's brand-name familiarity and a series of high-profile guests, the fledgling series has garnered solid ratings (Lloyd).

Original programming for the web is also on the rise, such as Netflix's *House of Cards* starring Kevin Spacey. And the trend seems to be increasing. CEO Reed Hastings plans to produce more original programming so Netflix isn't continually at the mercy of other content providers, noting that, "Ideally, we'll license two or three similar, but smaller, deals so we can gain confidence that whatever results we achieve are repeatable" (as quoted in Kafka). Describing caps and penalties imposed by the cable companies as "outrageous," Hastings clearly plans to make Netflix a content provider as well as a distributor—the old dream of controlling all aspects of the marketplace (Kafka).

Netflix's strategy seems to be working: as of April 2011, it had 23.6 million subscribers, or more than 7 percent of all Americans—more customers than the biggest cable TV network in the United States (Pepitone). Thus, even with the abortive attempt to separate streaming video and DVD rentals into two different services (quickly abandoned due to both impracticality and customer hostility), Netflix remains "the country's biggest provider of subscription video content" (Pepitone). And that number is only going to increase as Netflix positions itself to join Amazon's Kindle and Apple's iTunes as the chief streaming content providers in their respective areas. Yet surprisingly, Netflix is also boosting the sale of Blu-ray players and discs. If you have streaming video, why would you need a Blu-ray player? As Janko Roettgers discovered:

People don't necessarily buy Blu-ray players for the discs, but to access Netflix instead. However, once they take the players home, they apparently decide to buy a few discs and see what all the fuss is about. Despite the massive growth of Netflix, Blu-ray users are beginning to buy more discs than they did in the two previous years, according to NPD.

This means that Netflix may actually help Hollywood to sell Blu-ray discs by moving away from shipping discs itself. Of course, one big question remains: If people primarily buy Blu-ray players to access Netflix, why don't they just buy an Apple TV, Boxee Box or Roku device instead [to stream Netflix to their televisions]? NPD didn't explore this issue, but anecdotal evidence seems to suggest that people don't value media streaming devices the same way they do Blu-ray players. In other words, consumers don't see why they should spend $99 or more on a tiny little box if the same amount of money gets them a Blu-ray player that also streams online video.

Thus, there's no stable ground to stand on, if there ever was. One could easily argue that progress in the twentieth and twenty-first centuries has always moved at a rapid pace that is now accelerating and reaching the speed of light itself—because the landscape of both life and technology is constantly changing. The platform that seemed so solid yesterday will be obsolete tomorrow; the device "purposed" for one function will quickly be adapted for some other, seemingly competing one. And the rules of the game are changing. For example, the American Federation of Television and Radio Artists (AFTRA) negotiated the first contract with video game producers that guarantees the "principal performers" in streaming video games a chunk of the project's potential revenue, in the form of a 15 percent boost over the actors' normal fees for voicing key characters in the game (Evans).

Yet the sedentary gaming and web audience is only a small segment of the streaming viewer base. In the summer of 2011, the Virginia-based video platform company Voped announced the latest technological advance in its service: it can now offer smaller, independent content providers, whether producers or library sources, the "ability to live stream their free [or] pay per video content to any and all mobile devices, including the iPhone, iPad, and Android." Voped's president Mark Serrano notes that "multiplatform delivery is a critical and growing force within the online video industry," and for those seeking a distributor for their material, this advance puts handheld delivery within the range of nearly everyone (Bennett).

But the day of the truly independent content producer may be over. This is why YouTube introduced its 100 channels of "professional" programming and why the moving image industry as a whole continues to be dominated by the major studios. When the late Steve Jobs introduced Apple TV to the public in September 2010, he rhetorically asked, "So what have we learned from our users? They want Hollywood movies and TV shows whenever they want. It's not complicated. They don't want amateur hour. They want HD—everyone wants HD. They want to pay lower prices for content. They don't want a computer on their TV— they have computers. They go to their TVs for entertainment" (as quoted in Hosein). And that's why the day of the independent is dead and corporate visions are monopolizing the marketplace.

As the staff of the *International Business Times* wrote in a group report, mainstream providers like Netflix will continue to dominate streaming video downloads, as cable saturation dwindles: "As cable gets more expensive with more channels, customers increasingly want to pay only for what they want to watch. Netflix is an affordable option, and as more households gain devices to deliver streaming media to the family, the more the company secures its long-term profitability." And Netflix doesn't offer streaming videos of dancing cats or wacky skateboard accidents shot with a handheld cell phone. Netflix deals only in Hollywood movies—and some foreign theatrical films—that provide, above all, *entertainment* with glossy production values for viewers who simply want to "shut down" and watch, whether at home or on a plane or on vacation. With streaming video, you can carry your favorite movies anywhere, without having to worry about storing DVDs or "late return" fees—they're available at the touch of a button, day or night. This is what the twenty-first-century consumer of visual images wants, above all else: escapism.

Even though Apple is outwardly modest about the success of its Apple TV platform, which has thus far sold an impressive 4.2 million units, the company realizes that the future potential of its delivery system is virtually unlimited. Current Apple CEO Tim Cook has characterized Apple TV as "a hobby" for the company, commenting, "Our Apple TV product is doing quite well . . . but in the scheme of things, we still classify Apple TV as a hobby. We continue to add things to it. If you're using the latest one—I don't know about you, but I can't live without it. Other than that, no comment" (as quoted in Burns). Cook is obviously pleased with the progress Apple TV has made, and despite its technical imperfections, Apple TV is clearly the industry leader (for the moment) in streaming interconnectivity between conventional television and the web. Tech analyst Kasper Jade noted in late 2011: "Although Apple continues to see Internet television devices as a nascent category, frequently referring to the Apple TV as a 'hobby,' when sales of the device are [pitted] against its peers, the Apple TV appears to be a runaway success. For instance, Logitech said this week that 'very modest sales' of its $249 Google TV-based Revue set-top-box were exceeded by returns of the product from unhappy customers, prompting the company to slash pricing by 66% to match Apple TV's $99 price point."

But amid all the hyperbole about the dominance of streaming, a surprising fact emerges. According to Erik Gruenwedel, DVDs still form the backbone of Netflix's business model:

Streaming

On its way toward streaming dominance an interesting old school reality has emerged in Netflix's bottom line. DVD and Blu-ray disc rentals are delivering five times the profit margins compared with streaming. In dollars, that amounted to $194 million in profit for disc (52%), compared with $52 million (11%) for subscription video-on-demand (SVOD) in Netflix's most recent fourth quarter. The discrepancy underscores an inconvenient truth for Netflix. Namely that while the future may belong to streaming, the present still is very much a disc-driven business, no matter how much management wants to spin it otherwise. ("Disc Rental Profits")

Why is this true? Because a great deal of content simply isn't available as streaming video. The cost to license a film for video on demand—assuming it's a profitable, sought-after title—is much higher than the cost to license a DVD of the same title. And as streaming video becomes more popular, content providers are charging more for the rights to stream their films, convinced, as Gruenwedel rightly is, that streaming will become the dominant technology. But with a relatively small number of films available for streaming downloads even in late 2012, compared with the numerous titles available on DVD, Netflix is reluctantly stuck with the physical digital format for the foreseeable future and has only two choices in the end: stream everything, or lose its customer base.

Amazingly, Netflix often has only *one* DVD copy of less popular titles to service its entire rental base, so there's the problem of scheduling that single disc to satisfy multiple user requests and replacing it with a new one if it becomes damaged, which, given the uncertainties of DVD rentals, is inevitable. There are no such problems with streaming video; hundreds, even thousands of viewers can watch one master copy simultaneously. The convenience factor thus cuts two ways, for both consumer and supplier; DVDs may offer a greater selection (for the moment) and potentially better visual quality (presuming they're not damaged), but from an economic point of view, they've become impractical. And what will happen when a DVD goes out of print? What happens when it can't be replaced? Netflix is so anxious to be rid of DVDs that once a title enters the streaming universe, the physical DVD is summarily dropped from its inventory. Netflix is headed toward an all-streaming service, and the content factor has become almost incidental. If it can be streamed, Netflix wants it; if not, it doesn't.

That's why competition in this area is so brutal; indeed, Amazon On Demand offers many older titles as streaming downloads that Netflix ignores, and Redbox and Verizon are combining forces to launch a streaming video service to directly challenge Netflix, giving customers the option of either patronizing a Redbox kiosk or pushing a button and having content delivered immediately. With Redbox, the streaming option makes sense because each kiosk carries only 200 titles, on average, so streaming customers would have many more titles to choose from.

In May 2012 Amazon introduced what it calls the "Never Before on DVD" store, aimed directly at cineastes and buffs who are dissatisfied with Netflix's mainstream offerings. Like the vinyl LP, which is now enjoying a remarkable resurgence among audio aficionados, the DVD still has an aura of quality and relative permanence that streaming, by definition, can't duplicate. As Marc Graser reported in *Variety*, "Amazon is offering more than 2,000 titles from the homevid divisions of Disney, Sony, Warner Bros., Lionsgate, Universal and 20th Century Fox. It will also offer up films from MGM, and CBS Networks. MTV and Nickelodeon are offering TV shows." Amazon wouldn't embark on such an enterprise if it couldn't make money on it, nor would the studios participating in the venture. So it seems there's still at least a niche market for DVDs.

But there's an even more interesting platform content on the horizon. According to Eric Jackson, it's possible that the entire infrastructure of the web will collapse and be replaced by a whole new way of interfacing with online content; in his words, "Google and Facebook might completely disappear in the next five years." As Nick Foley reports, "According to market research firm YPulse, 18% of teens prefer to 'check in' on Foursquare instead of Facebook, and 10% say Pinterest is a better site for browsing. More than eight years after Facebook's inception, its mass appeal has drawn older crowds who add their kids as Facebook friends. That development could be tarnishing the site's 'cool factor' in the eyes of teens, said Jake Katz, chief architect at YPulse" ("Teens Turn from Facebook"). The concept of Facebook vanishing from the social media horizon isn't as far-fetched as it sounds. People seem to live in the eternal present, convinced that because something exists now, it will almost inevitably continue to exist in the future. But our own personal mortality should instinctively tell us otherwise; nobody lives forever, and when you're on top, the only way to go is down, unless you get bigger and bigger and swallow up the competition. But consider MySpace: what was once the most vibrant social networking site on the web is now a virtual graveyard. Its value has fallen to nearly nothing since Facebook supplanted it as the "go to" social destination. Search engines like Ask Jeeves and Kartoo have all but disappeared, and despite its promise to protect user privacy and a high-profile conventional television campaign—a rather desperate attempt to reach an audience that doesn't *watch* TV anymore—search engine Bing also failed to gain significant traction.

Hand in hand with this development is the unlikely suggestion from Facebook and other social media giants that their employees actually "step back" from their online lives a bit and experience the real world. As Matt Richtel reported on July 23, 2012:

> Stuart Crabb, a director in the executive offices of Facebook, naturally likes to extol the extraordinary benefits of computers and smartphones. But like a growing number of technology leaders, he offers a warning: log off once in a while, and put them down. In a place where technology is seen as an all-powerful answer, it is increasingly being seen as too powerful, even addictive. . . . "We're done with this honeymoon phase and now we're in this phase that says, 'Wow, what have we done?'" said Soren Gordhamer, who organizes Wisdom 2.0, an annual conference he started in 2010 about the pursuit of balance in the digital age. "It doesn't mean what we've done is bad. There's no blame. But there is a turning of the page." . . . Richard Fernandez, an executive coach at Google and one of the leaders of the mindfulness movement, said the risks of being overly engaged with devices were immense. . . . "It's about creating space, because otherwise we can be swept away by our technologies."

Indeed, the risk of being buried by an avalanche of material is enormous, because no matter how much you see on the web, there's always a link that takes you someplace else and then someplace else after that and someplace else after that. The problem is that it's almost impossible to find one's center, since the web itself extends endlessly in all directions, with no beginning or end. It has also become impossible to ignore or even to avoid the web; with telephone books, directory assistance, and human help lines being phased out, one is forced to embrace the digital world.

Streaming

It appears that text in any form is becoming outdated, as streaming images and spoken words replace existing databases. Qwiki, a new automated slide-show display that takes you through the basic outlines of someone's life with a few well-chosen images, is an excellent example of this emerging technology. The interesting thing about Qwiki is that no human agency seems to be involved; it's simply a program that trawls the web, picks up relevant text and images, and puts them together in a highly interactive presentation. A synthetic voice reads the text in the background as the words appear at the bottom of the screen, as if subtitling the presentation. As Sarah Yin notes, "A search on Qwiki pulls up a topical, Wikipedia-like page with a 'rich media narrative' of videos, photos, and audio clips relating to the topic. . . . Take, for example, a Qwiki search on Jack LaLanne, the [pioneering] bodybuilder. . . . The entry provides a quick, minute-long audio-visual entry about who LaLanne is, complemented by a four-page slideshow of photos of LaLanne and even of his parents' hometown. At the end, you are given options to share the entry through various social networks."

Support for the site is surprisingly person-to-person; I've had contact with several of Qwiki's content developers and suggested how they could better mine Wikipedia for more images, better text, and more trenchant video. Numerous other "encyclopedia" sites on the web utilize Wikipedia's research (which can be copied and used by anyone) as the basis for their own products, but with Qwiki, there's really something new and innovative going on—it takes the information available and makes it more directly accessible to the public. And with the automated voice–text recognition system, even the visually impaired find the site useful, as it reads the text aloud for all to hear. Qwikis are rapidly becoming ubiquitous, covering nearly every topic on Wikipedia. Users of the site are also invited to create their own Qwikis.

It's interesting that Qwiki's offices are located in Manhattan rather than in Silicon Valley. As Georgia Kral discovered, the project was initiated in California in 2009, but after three years on the West Coast, founder and CEO Doug Imbruce, a native New Yorker, decided that due to various tax incentives and other inducements, New York City was the ideal place for Qwiki to grow and thrive in the future. As Imbruce said of the move, somewhat chauvinistically, "Qwiki is a tool for creative people, and New York's the epicenter of creativity worldwide." Right after the move to Manhattan, Qwiki struck a deal with ABC News to create Qwikis based on its programming, to be used as part of ABC's online presence. Qwiki's new home is located right next to ABC's corporate headquarters, and that proximity clearly influenced the speed with which the deal was consummated. Without the New York City location, "who knows if it would have happened?" And even if such an alliance had eventually been forged, "it wouldn't have happened as quickly," Imbruce stated (as quoted in Kral).

There seems to be no end to the innovation engendered by such rapid expansion and visionary thinking. The web, it can safely be said, will remain in a constant state of "becoming," never staying the same for long, before the next new wrinkle emerges. As Eric Jackson points out: "The bottom line is that the next 5–8 years could be incredibly dynamic. It's possible that both Google and Facebook could be shells of their current selves—or gone entirely. . . . It's a lot easier to start asking Siri [Apple's voice-activated mobile application] for information instead of typing search terms into a box. . . . In all likelihood, we could have an entirely new way of gathering information and interacting with ads in a new mobile world than what we're currently used to today."

Siri is indeed another game-changer. Apple launched Siri with a series of subdued television spots featuring John Malkovich conversing with Siri as if "she" were a sentient being rather than an automated phone application. This is what Apple wants you to think. Siri can be a companion, a helper, a guide through the web. Unlike previous search engines, all you have to do is ask Siri a question. And that's not all: if Siri needs more information to complete a task, it asks you directly, as a real person would. Siri represents an entirely new way to browse the many functions of the web, and in many ways, it is (for the moment) the ultimate streaming device; it streams your conscious desires as you speak them out loud. As Apple describes it, "You can speak to Siri as you would to a person—in a natural voice with a conversational tone. If you want to know what the weather will be like tomorrow, simply say 'What will the weather be like tomorrow?' or 'Does it look like rain tomorrow?' Or even 'Will I need an umbrella tomorrow?' No matter how you ask, Siri will tell you the forecast. . . . And the more you use Siri, the better it will understand you. It does this by learning about your accent and other characteristics of your voice" ("Siri—Frequently Asked Questions"). Siri can also learn about the "key relationships" in your life and store that information, asking you, for example, the names of your siblings so Siri can recognize them later and put them in context with the social fabric of your existence. Siri takes dictation and can understand and respond in American English, British English, Australian English, French, German, Japanese, Chinese, Korean, Italian, and Spanish, with more languages to come. Siri, after all, is in the "beta" or testing stage as I write this. But it won't remain there for long, except in the sense that it will continually be evolving.

Siri is designed to be your friend, ally, and confidant, as well as maintaining your social calendar. As demonstrated in the television spots, Siri's soothing tone can be a source of warmth and comfort; Siri takes care of you, almost anticipating your needs. Who would have imagined a device of such depth and flexibility even a year ago, much less five? Once upon a time, you typed in a search term, and the information appeared onscreen. But now, even that's too much work. Siri will guide you through all that's available on the web and, if you so desire, stream it to you in the form of music, movies, television shows, or text. All you have to do is ask. You can send e-mail by dictation—no more thumbing through your Blackberry keypad. You can search the web, use the phone, find out how your stocks are doing, and even browse Wikipedia, all through voice commands.

Given the current capabilities of Siri, it is clear that Eric Jackson's call for the embrace of constant mutability is simply common sense. Some observers, however, such as Nick Bilton (who is always on the cutting edge of emerging technology), have "soured" on Siri, claiming that it isn't sufficiently responsive to voice commands; they prefer to use Google Voice Search to trawl the web. According to Bilton, Siri doesn't work as well in public spaces as it does in private. In a July 15, 2012, article, Bilton cited the work of researcher Gene Munster, who "subjected Siri to over 1,600 voice tests, half in a quiet room and half on a busy Minneapolis street. In the quiet room, Siri understood requests 89 percent of the time, but she was able to accurately answer a question only 68 percent of the time. On a busy street, Siri could comprehend what people were saying 83 percent of the time, but answer a question correctly only 62 percent of the time." But Siri is still in the rollout phase; with time and increasing refinement, Siri or another voice command–voice recognition service like it will undoubtedly improve significantly and be a key component of the Internet of the future. The platforms of the web, voice, text, image, and sound are constantly changing, and each genuinely new paradigm will render all previous platforms obsolete. It's technology, it's progress, it's innovation—it's life in the twenty-first century, moving at the speed of thought.

Soon, we may move beyond the articulation of thought to brain impulses themselves. Already, a television remote-control device called the Haier Brain Wave TV has been unveiled: it uses the power of a viewer's mind to control the family television. Julie Lasky reports, "As demonstrated at the Consumer Electronics Show in January [2012], viewers slip on a headset developed by the company NeuroSky, specialists in 'brain-computer interface technologies,' and work to translate bursts of cerebral electricity into operating commands." The technology is in its infancy, of course, but the most important point is that it exists and that the boundaries of what's possible are being pushed further and further out to the perimeters of functional reality.

Nor is this the only gadget that has made the supposedly impossible an accomplished fact. With DigiMemo, you write on a digital notepad, and your text and drawings are instantly uploaded to your Mac or PC without touching a keyboard. There's a new "video pen" with an eight-gigabyte memory that can record sixty minutes of high-resolution video at the touch of a button, or 30,000 still pictures, and it costs less than $200 retail. The Looxcie "wear and share" video camera fits over your ear like a mobile phone device but functions as a point-of-view (POV) video recorder, documenting your life as you live it. The device stores up to five hours of uninterrupted video that can be uploaded to your computer or Facebook or live-streamed on the web as it happens, turning the wearer's existence into a global video stream. The Looxcie, which weighs less than one ounce and is about three inches long, is relatively unobtrusive, yet it can broadcast what you see to every corner of the planet. "Share your life as you live it," Looxcie's promotional materials urge. You can "view and control video on your smartphone, share clips instantly with friends and family on Facebook, YouTube, Twitter, or as e-mail attachments, or go live with real time personal stream-casting."

And the price for all this? Just $149.99 retail. It isn't an exaggeration to say that such a device will alter the world as we know it, change the rules of documentary video production, and, if it catches on, inundate the web with millions of hours of personal data, all of which can be scanned, indexed, and subjected to facial recognition technology, exposing everyone in its range to a degree of hypersurveillance hitherto thought impossible. Looxcie, of course, views the prospect of personal POV video streams as a utopian social horizon, promising that "Looxcie sees what you see, and records your life as you live it, from the everyday to the most amazing adventure. When something special happens, Looxcie can automatically send your video to your social network or email. You don't even have to go back to your computer to share your videos. Just press a button on your Looxcie camera and immediately upload the video to your favorite social media site" ("Looxcie 2").

The Echo Smartpen functions along the same lines, although it positions itself as a learning tool for the academic setting, as a study aid for students and professionals. Once again, the device functions as an omnivorous harvester of data, but the Smartpen can record up to 800 hours of *audio* lectures, discussions, meetings, and the like, while appearing to be an ordinary pen, such as one would use to take conventional notes. But as its advertising materials point out, the Smartpen does much more than merely record audio or serve as a writing tool: "Smartpens record everything you hear, write and draw so you'll never miss a word. Replay your meetings or lectures simply by tapping on your notes. Save, search and organize your notes on the Mac or PC for fast, easy access to what's important . . . easily send notes and audio to people and destinations [via e-mail, and] convert your handwritten notes into digital text." Some of these functions seem useful and practical, but on the whole, the Smartpen is an invasive product, a recording and surveillance device masquerading as a conventional pen. Essentially, everything the Smartpen records can be instantaneously uploaded to the web as streaming text, possibly making even the most intimate conversation a global information stream. The potential abuses of such a device are myriad, and they will no doubt manifest themselves in the future, because the technology involved is both affordable and readily available.

Everyone can afford a Looxcie or a Smartpen, and with them, everyone can broadcast their "lives" to the world at large. How will all this information be used? How will it be stored? How will governments, corporations, individuals, and other entities use or misuse the cornucopia of information these devices provide?

With all these new technological advances, there's no way of putting the genie back in the bottle. In the twenty-first century, personal privacy seems an utterly outdated concept. All of us are potential sources of live streaming video and audio, sent out into the ether to be witnessed by anyone who happens to be watching. In some ways, this will level the social playing field; with Looxcie, for example, anyone can become a documentarian, and if the device is used to uncover highly volatile, unstable, or corrupt social environments, the information provided could potentially topple a totalitarian regime or expose unethical business practices. But somehow, I think most of the video recorded on Looxcie will wind up on *TMZ* or some other gossip site, disseminating embarrassing or even "career-ending" moments with tabloid relish. The new world of social media is, as many have observed, absolutely open to constant permutation. Who knows what impact these devices will ultimately have on our culture?

As we've seen, social media have become an increasingly important part of the conventional television landscape. Nina

Tassler, president of entertainment for CBS, told advertisers at a public meeting in 2011, "We've amassed 100 million fans on social media . . . that's one hell of a water cooler" (Stelter and Carter). With so many viewers linked in to CBS through Facebook and other social media sites, the next logical step is to extend that connection to advertisers, which, after all, are less concerned with the content of programming than with the demographics of the audience and the number of viewers watching at any given time. A relatively new mobile app, Shopkick, hopes to capitalize on the social media aspect of network television programming by offering special discounts to viewers who open up the app at a certain time, such as when a store has a commercial running. "As Cyriac Roeding, chief executive of Shopkick, put it: 'The cellphone is the only interactive medium that you carry with you while you're watching TV *and* while you're shopping in the store. The cellphone is therefore the only interactive medium that can function as the bridge between the TV screen and the store shelf'" (Stelter and Carter).

But into this rather pervasive advertising landscape, a new element has been introduced: Dish Network's Auto Hop DVR, which can "automatically recognize and skip over ads or recorded shows" (Lieberman). Needless to say, this idea doesn't appeal to advertisers or to network CEOs—either conventional or cable. David Zaslav, head of the Discovery Channel, proclaimed that the Auto Hop DVR "could create real carnage for the industry" (Lieberman). As described by Shalini Ramachandran, "With Auto Hop, viewers see a black screen momentarily where the ads were broadcast, or a glimpse of the first frame of the first commercial. Then the show resumes. Consumers merely have to click an on-screen Auto Hop button before a show to enable the feature." According to Tracey Scheppach, innovations director at Starcom MediaVest (a media-buying firm owned by Publicis Groupe SA), "There has been a problem with ad skipping and this is just making it worse." But Dish chief executive Joe Clayton counterclaimed that "the Auto Hop feature is all about the consumer. This has been the Holy Grail of television viewers for 40 years. What's wrong with giving the consumer what he wants? That's my response to anybody who takes issue with this" (as quoted in Ramachandran). Conventional DVRs have been "widely adopted and are now in about 43% of U.S. households, according to Nielsen. Media buyers say about 50% of ads get skipped by DVR users" (Ramachandran). With Auto Hop, the percentage of ads skipped seems very likely to increase.

Whereas Tim Armstrong, the head of AOL, suggested rather disingenuously that "Dish's initiative puts more pressure on advertisers and media companies to develop commercials that people will want to watch" (Lieberman), Fox, CBS, NBC, ABC, and other broadcast and cable content providers reacted vociferously to this new technological development that, for once, favors the consumer and not the media conglomerate. For its part, Fox described Auto Hop as "a bootleg broadcast video-on-demand service" and argued that Dish "is undermining legitimate consumer choice by undercutting authorized on-demand services and by offering a service, that, if not enjoined, will ultimately destroy the advertising-supported ecosystems that provide consumers with the choice to enjoy free over-the-air, varied, high-quality primetime broadcast programming" (as quoted in Hachman).

NBC Universal's spokesperson said, "Dish simply does not have the authority to tamper with the ads from broadcast replays on a wholesale basis for its own economic and commercial advantage," while CBS added, "This service takes existing network content and modifies it in a manner that is unauthorized and illegal. We believe this is a clear violation of copyright law and we intend to stop it" (as quoted in Nakashima). This is a battle with far-reaching consequences for all concerned. Once again, though, the genie can't be put back in the bottle. The very fact that Dish introduced Auto Hop, knowing full well that advertisers and content providers would object and would almost certainly take legal action (ABC, NBC, CBS, and Fox have all sued Dish, and Dish has filed a countersuit of its own), suggests that the idea has consumer traction, giving viewers *some* control over the advertisements that clutter commercial television.

Nevertheless, people still watch a hell of a lot of conventional broadcast television, despite the nationwide switch to digital broadcasting and the cessation of analog transmissions, which had been the industry standard since the medium's inception. As the Nielsen Company reported in June 2011:

Those who view 18.8 minutes of streaming video a day also watch 272.4 minutes of traditional TV a day. The lightest streamers, those viewing 0.1 minutes of streaming video, watch 290.0 traditional TV minutes a day. In the fourth quarter of 2010, the heaviest streamers were at 14.5 minutes of daily streaming, and 262.7 minutes of traditional TV . . . while certain segments of the population are migrating toward specific services and viewing habits, the resounding trend is this: Americans are spending more time watching video content on traditional TV, mobile devices and via the Internet than ever before. (Friedman)

Dish Network's Auto Hop offers a practical solution to something that nearly everyone agrees on: there are too many commercials in the media. (Even the lower-end Kindles force readers to view pre-text commercials while ostensibly reading a book—something I doubt Charles Dickens or Edgar Allan Poe had in mind, though I'm sure that ultracommercial author James Patterson wouldn't object.) Whether Auto Hop ultimately succeeds or not, it seems that in those "158 hours and 47 minutes" of television that people watch each month, there are an awful lot of advertisements they would rather not see. So it's possible that with Auto Hop or something like it, the contours of the conventional ad-supported television platform could shift decisively, perhaps toward a pay subscription model that is free of advertising.

There's also Aereo, which streams regular television programming directly to the web. Media critic Stephen Battaglio describes it as "a new service that enables subscribers to watch and record broadcast TV signals over the Internet for only $12 a month. . . . Aereo's stance is that its product is a new technology (a dime-size antenna is assigned to each subscriber) that enables online users to get signals that are already free" (7). As one might expect, Aereo has encountered significant legal opposition from conventional television networks and content providers, but thus far, the service, backed by billionaire entrepreneur Barry Diller, has survived a major court battle, and it seems on track to roll out on a national basis. As Edmund Lee and Jonathan Erlichman reported on July 12, 2012:

> A U.S. district judge this week allowed Aereo to continue operating while television networks pursue a copyright lawsuit against the company. Aereo captures broadcast signals with small antennas and streams them to devices such as Apple's iPad, without paying for the programming. "The ability for consumers to receive broadcast over the air signal is their right," Diller said. The ruling could upend the economics of broadcast television. CBS, NBC, Fox and ABC—all plaintiffs to the lawsuit—receive "retransmission consent fees" from pay-TV operators such as Time Warner

Cable for the right to rebroadcast the free-to-air signals to their subscribers [but not from Aereo]. "I did think we were on the right side of this, and I'm happy the judge agreed with us. We're going to really start marketing [now]. Within a year and a half, certainly by [20]13, we'll be in most major [markets]," [Diller added].

In the digital era, change is happening in every platform: TV, video, music, and especially the news media. Newspapers are becoming obsolete. John Paton, head of the Media News Group, which controls a number of publications including the *Denver Post, Detroit News, Salt Lake Tribune*, and *San Jose Mercury News*, has been aggressively pushing for the complete digitization of all the journals under his command. According to Paton, the newsprint industry needs to "stop listening to newspaper people" in order to survive (Carr, "Newspaper's Digital Apostle," B1). Books, as we've seen, are read on Kindles or Nooks, and most music is now sold and streamed through iTunes and Amazon. Even the venerable magazine *Newsweek* seems headed for a digital-only future, despite editor Tina Brown's denials. *Newsweek*'s owner, Barry Diller, has commented, "I'm not saying it will happen totally, but the transition to online from hard print will take place. We're examining all of our options" (as quoted in Dolor). But what other "options" are there? It's just like the movies: film is dead; digital production rules. So will it be with print publications.

More people now stream their newspapers online than read print editions, and who can blame them? Streaming news delivers information faster and with greater timeliness, and stories can be updated with a few keystrokes to keep abreast of breaking news. Some magazines sell more copies on Apple's newsstand app than they do on real newsstands (Chozick), with *Time* and its sister publications *People, Sports Illustrated, Entertainment Week-ly,* and others hopping on the virtual bandwagon. As new apps, such as "Viddy, a social media app for posting and sharing video; Frameographer, for recording time-lapse and stop-motion video; Carcassonne, a strategy game for iPad; and Angry Birds Space, the popular game that appeals to adults and children" (Eaton), continue to proliferate, it's clear that the culture of both news and entertainment has joined the streaming parade—along with its related ad content. What were once print ads are now iPad advertisements; the era of print, like that of celluloid film, VHS tapes, CDs, DVDs, and other physical formats, has passed. It's all streaming, all the time.

Another platform that's changing is the distribution model for performers seeking to disseminate their work directly to their fans and admirers without going through a conventional media outlet. Writer, producer, director, editor, and actor Louis C. K. recently put this into practice after building a solid fan base through his edgy sitcom *Louie* on the FX network, which was recently renewed for its third season and garnered a host of Emmy nominations along the way. Like all genuinely innovative artists, Louis C. K. has sought to control every aspect of his work throughout his career. He only recently turned over the editing of *Louie* to Woody Allen's former cutter, the gifted Susan E. Morse, but the results are electric. *Louie* is unlike anything else on television. After a twenty-five-year slog to the top, Louis C. K. has emerged as perhaps the most influential and savage social satirist of the early twenty-first century, in the tradition of Mort Sahl and Lenny Bruce.

Louis C. K. has done numerous stand-up specials in the past, but given the extremely controversial material and strong language, they all wound up on HBO, Showtime, and other "premium" cable services, partly because of censorship concerns, but also because, historically, that was the most successful distribution model. Reruns of Louis C. K.'s specials on *Comedy Central*, for example, are "necessarily" censored, diluting the impact of his material. So, for his new project, Louis C. K. decided to distribute an hour-long stand-up concert via the web as streaming video, at a cost of $5 per download. The final show was entitled *Live at the Beacon Theater*, and against the advice of everyone, Louis C. K. put it up on his own website, free of encryption or other copy-guard restrictions, and waited to see what happened. As he reported, with his customary bluntness, in a mass e-mail to his fans, the experiment was a surprising success:

What I didn't expect when I started this was that people would not only take part in this experiment, they would be invested in it and it would be important to them. It's been amazing to see people in large numbers advocating this idea. So I think it's only fair that you get to know the results. . . . The show went on sale at noon on Saturday, December 10th [2011]. 12 hours later, we had over 50,000 purchases and had earned $250,000, breaking even on the cost of production and website. As of [December 14, 2011] we've sold over 110,000 copies for a total of over $500,000. Minus some money for PayPal charges etc, I have a profit around $200,000. . . . If the trend continues with sales on this video, my goal is that I can reach the point where when I sell anything, be it videos, CDs or tickets to my tours, I'll do it here and I'll continue to follow the model of keeping my price as far down as possible, not overmarketing to you, keeping as few people between you and me as possible in the transaction.

In the end, this may be the most important and influential platform shift of all: performers, writers, directors, and artists controlling their own material from start to finish, marketing it themselves through their own websites, charging a reasonable price for it, and cutting out the middleman entirely.

Movies, television, books, and music are all creative arts; they're inextricably tied to commerce, whether they want to embrace it or not. Films, in particular, need to make a lot of money to earn back their production and promotion costs, and while the need for programming is certainly evident, the concept of producing new material is becoming more and more complex and fraught with financial uncertainties. There has always been risk attached to any film or television project, of course, but in the twenty-first century, both above-the-line costs (for talent) and below-the-line costs (for production) have increased so dramatically that one badly placed "bet" on a big film can sink an entire studio. As Adam Davidson recently mused, "How does the film industry actually make money?" He notes:

I've been trying to come to terms with two seemingly ir-reconcilable facts. First, *Men in Black 3* has made more than $550 million worldwide. Second, while a representative from the parent company of Columbia Pictures told me that the movie is now "in the win column," it seemed until recently as if Columbia might actually *lose* money on it. How could that be? It's not so complicated. Its production costs were close to $250 million; worldwide marketing most likely added at least that much; and a big chunk of the ticket sales go to theaters and distributors. There must be an easier way to make money. For the cost of *Men in Black 3*, for instance, the studio could have become one of the world's largest ven-ture-capital funds, thereby owning a piece of hundreds of promising start-ups. Instead, it purchased the rights to a piece of intellectual property, paid a fortune for a big star and has no definitive idea why its movie didn't make a huge profit. Why is anyone in the film industry?

It's a very good question. As Davidson summarizes: "1. Hollywood requires more blind guesswork than most industries; 2. Still, the model remains largely unchanged; 3. That's partly what keeps innovators away" and keeps sequels in the pipeline. But even with such a resolutely presold commodity as *Men in Black 3*, the profit margin is relatively slim. Davidson observes:

> The reason a majority of movie studios still turn a profit most years is that they have found ways to, as they say, monetize the ancillary stream by selling pay-TV and overseas rights, creating tie-in video games, amusement-park rides and so forth. And the big hits, rare as they may be, pay for a lot of flops. A modern studio's main asset, however, is its ability to put together these disparate elements. They know how to get Tom Cruise to do a film, how to get it into theaters around the country and whom to call to set up a junket in Doha. They also know the industry's language of power, with its ever-changing rules about which stars, restaurants and scripts are cool and which are not. It's the stuff of easy parody, but it's worth billions.

But this bottom-line mentality stifles any real innovation and discourages even the slightest element of risk, creative or otherwise. One can't continue making sequels forever; eventually, a new model has to be created, if only to serve as a basis for future variations on a theme. And so, as always, the only really new work comes from the edges of the industry and drives it forward to create new formats, new narratives, and new stars. While corporations and conglomerates seek to "monetize" all the content they own, many members of the creative community simply want a fair return on their investment, with as little front-office interference as possible. And in this light, Louis C. K.'s model seems like an admirable alternative to conventional media distribution systems.

Streaming video, text, music, and information doesn't have to be controlled by the few; this was the impetus behind the creation of the web in the first place: the free and egalitarian exchange of visions. As every new platform emerges, it can expect both user feedback and pushback, and in the end, no one needs to join a social networking site to be connected to the rest of the "webiverse." You're already connected, already out there, part of the vast array of information and entertainment available for instant download, and whether you pursue the corporate or the personal model, the ultimate platform is one that *you create.* Every other attempt at mediated self-expression comes with strings of various kinds attached; only when you assume control over your identity and your work can a sustainable platform truly be created, one that will continually evolve along with its creator.

5

Streaming the World

As streaming technology progresses, the vast amount of information that each of us possesses will become more and more available to even the casual observer, who won't even have to search for it. No search engines, using either keyboards or voice commands, will be necessary. Simply walk around the streets, and your entire history will soon be on display. The reason: Google glasses. Whether one regards Google glasses as invasive or not is almost beside the point; like the other technological advances discussed in this volume, they exist and will inevitably be utilized. Basically, Google glasses turn every person, building, automobile—any object of any kind—into an instant font of information. For people, this information includes name, personal history, marital status, income, job, and other available data. As Nick Bilton reported on February 22, 2012:

> By the end of 2012, Google is expected to start selling eyeglasses that will project information, entertainment and, this being a Google product, advertisements onto the lenses. The glasses are not being designed to be worn constantly —although Google engineers expect some users will wear them a lot—but will be more like smartphones, used when

needed, with the lenses serving as a kind of see-through computer monitor. . . . Through the built-in camera on the glasses, Google will be able to stream images to its rack computers and return augmented reality information to the person wearing them. For instance, a person looking at a landmark could see detailed historical information and comments about it left by friends. If facial recognition software becomes accurate enough, the glasses could remind a wearer of when and how he met the vaguely familiar person standing in front of him at a party. . . . When someone is meeting a person for the first time, for example, Google could hypothetically match the person's face and tell people how many friends they share in common on social networks.

Even for self-confessed tech freaks, Google glasses seem a bit over the top, but that's just for now. When we all have the capability to view the world from a privileged point of view with a wealth of superimposed information, how can we say no? People who wear Google goggles will have an immediate upper hand in almost any situation, and anonymity will completely evaporate. As Damon Brown points out:

Early reports say that the glasses aren't designed for everyday wear, . . . [but] that's akin to saying that smartphones aren't meant to be carried all day or home computers were only for certain tasks—before we carried our smartphones all day and used home computers for most tasks. I doubt that Apple planned on people texting while walking, either. [The] glasses are actually the final piece to Google's mission; to know what a user [is] doing every single moment of the day. The search giant already is unifying some 60-odd products into one log-in for continuous online tracking. . . . [When] wearing Android-powered glasses, you're giving

Google unprecedented access to your location at all times, your most common interactions, your closest companions through facial recognition, and your eating, shopping, and traveling habits.

The glasses will cost somewhere between $250 and $600, depending on the features they incorporate. There will be "basic" Google glasses as well as more sophisticated models, as with every other electronic device. They will turn the wearer not only into a *source* of information for both Google and other wearers of Google glasses but also a constant *target* of a stream of ads based on the user's location, age, gender, and known product preferences. As Brown notes:

It's all about location, location, location, and Google's goggles will have a direct bead on you 24 hours a day. On one end are Groupon, Living Social, and other mass-coupon services, and on the other are FourSquare, Gowalla, and other check-in companies. Google jumped into the middle of the fight [in 2011] with its Google Coupons app and, more recently, with Google Latitude check-ins. Both Google Coupons and Google Latitude aren't making much of a dent in the competition, but having tracking data on users 24/7 would be a huge coup for both services. It would be automatic check-ins, pushed suggestions, and coupons. . . . We've seen accident after accident of people texting, gaming, or web surfing while walking. The U.S. government is now considering banning all automobile phone calls, including hands free. Doesn't it seem like a bad time to develop digital displays *in front of our eyes?*

Perhaps, but as of June 13, 2012, there was already a promotional video for Google glasses online. Thus far, it has had more than

16 million views, meaning that (1) a large portion of the public is waiting to embrace the glasses, and (2) they'll soon be available. Google cofounder Sergey Brin was spotted wearing a pair of the glasses at a party in San Francisco in April 2012, and tech pundit Robert Scoble snapped a picture of Brin with actor Philip Seymour Hoffman, who was also wearing the glasses. Tellingly, Brin wouldn't let Scoble look through the glasses himself, "but I could see they were flashing info to him," Scoble stated (as quoted in Ionescu).

The folksy, low-key, point-of-view promo video teasingly entitled "One Day . . ." shows someone waking up with the glasses on, eating breakfast, walking over to the Strand Bookstore in New York, meeting with a friend, snapping digital photos of his surroundings, listening to music through the glasses' audio system, video-conferencing with his girlfriend, and generally existing in an utterly plugged-in world. He seems to be entirely comfortable with all this, and guess what? If you have prescription glasses, you can get Google glasses made to order to fit your needs (Franzen). There is little doubt that they will become "must have" items for the tech set. Soon you'll be able to see for yourself what the world looks like laid out in a grid— a grid where Google knows your location 24/7 and can sell you anything it wants to, all in the name of convenience, of course. The glasses will be voice controlled, so you will be able to summon up phone or video connections, recording capability, facial recognition software, street directions, restaurant reviews, and the location of your friends, all with vocal commands.

Howard Baldwin and Ed Oswald of *PC World* had an interesting discussion on the pros and cons of Google glasses in early April 2012. Baldwin lauded the idea, noting:

People have been trying to build wearable computers for years. Project Glass [Google's in-house name for the project] puts the technology into something people already wear. . . . Who doesn't love hands-free computing? Maybe these will help us bypass those nanny-state laws and let us talk while we're driving again. . . . Now that computers will be on your head instead of in your pocket or purse, you can identify other technophiles more readily on the street. . . . Mobile anytime "knowledge" will be breakthrough technology for handicapped people of all types, especially for intellectually and physically challenged. This technology could also revolutionize knowledge, manufacturing, and service industries allowing for fast access to needed data, maps, and schematics.

Oswald countered:

Using the glasses will likely be more distracting than texting currently is. Google glasses places the data in front of your line of sight so that you probably will focus on the data rather than what's around you. This could be more dangerous than texting or using your cell phone while driving. [There are also] potential privacy issues. As we wear these glasses around town, the search giant might be able to gather even more data on our daily lives. The [promotional] video clearly shows deep integration to Google services: you are encouraged to share with the search giant. You think Google's ads are too personal now? Imagine those ads after wearing Google glasses!

Yes, one can only imagine.

But of the hundreds of comments posted in response to this discussion, most were overwhelmingly positive. One reader observed, "If Google glasses supported both TV and web browsing, it could allow TV commercials to link to the shopping and purchase page. Plus then video, games and shopping could all be in 3D!" Another enthused, "What would make Google glasses a hit is if it supported all previous mediums, hyperlinking them all together, and in 3D (TV, books, games and apps, web browsing, shopping)." A third reader commented, "If it works as well as the video, I'll give it a shot" (as quoted in Baldwin and Oswald).

"My son is four years old, and this is going to be his generation's reality," adds Guy Bailey, a social media theorist based in Atlanta, Georgia. Bailey even expects that Google glasses will ultimately be replaced by "body implants, so that in 10 years or so you'll be able to get a 'heads up display' inside your head" (as quoted in Ortutay, A9). "There is a lot of data about the world that would be great if more people had access to as they are walking down the street," agrees Jason Tester of the Institute for the Future in Palo Alto, California. But Tester cautions that "once that information is not only at our fingertips but literally in our field of view, it may become too much" (as quoted in Ortutay, A9). Barbara Ortutay adds, "It doesn't take much to imagine the possibilities. What if you could instantly see the Facebook profile of the person sitting next to you on the bus? Read the ingredient list and calorie count of a sandwich by looking at it? Snap a photo with a blink? Look through your wall to find out where electrical leads are, so you know where to drill?" (A9). Certainly, Google is thrilled with the concept, and CEO Larry Page argues that "things we used to think were magic, we now take for granted: the ability to get a map instantly, to find information quickly and easily, to choose any video from millions on YouTube rather than just a few TV channels" (as quoted in Ortutay, A9).

Google glasses, clearly, are just the next step in this evolution. "This puts Google out in front of Apple; they are a long ways ahead at this point," notes Michael Liebhold of the Institute for the Future. "In addition to having a superstar team of scientists who specialize in wearable [technology], they also have all the needed data elements, including Google maps. . . . Of course, it could be really annoying, but if it's handled well, it could be a nice complement to reality" (as quoted in Bilton, "Rose-Colored View"). It's the "hands-free" aspect that apparently appeals most. As one prototype user points out, the glasses "let technology get out of your way. If I want to take a picture, I don't have to reach into my pocket and take out my phone; I just press a button at the top of the glasses, and that's it" (as quoted in Bilton, "Rose-Colored View"). Since that prototype was introduced, the "button" function has been eliminated; as the promotional video makes clear, it's all done with voice commands. In late June 2012 Google staged a "spectacular" stunt involving Google glasses, a group of daredevil skydivers, and live video, ensuring a great deal of positive media coverage, at least within the tech community (Orlin). It's all razzle-dazzle, of course, but then again, it's effective advertising. When the Google glasses are finally rolled out, there's no doubt they'll find numerous adherents who can't wait to become even more enmeshed in the digital universe.

And yet there are upsides to this new "instant read" visual technology. One of the most useful is the Word Lens translation app, which can be added to your iPhone camera function and seamlessly translates written English into Spanish, or vice versa. As Charlie Sorrel wrote of a test run of the Word Lens in December 2010:

It's an augmented-reality, OCR [optical character recognition]-capable translation app, but that's a poor description. A better one would be "magic." Word Lens looks at any printed text through the iPhone's camera, reads it, translates between Spanish and English. That's pretty impressive already—it does it in real time—but it also matches the color, font and perspective of the text, and remaps it onto the image. It's as if the world itself has been translated. Impressed? You're not the only one. John Gruber of *Daring Fireball* puts it best: "[It's] as though near-future time travelers started sending us apps instead of Terminators." Word Lens is a taste of science fiction [but] instead of being a convenient device to avoid movie subtitles, it's a real, functioning tool.

Of course, with the introduction of Google glasses, Word Lens seems instantly outdated; however, it will no doubt find a new home as an optional feature of Google glasses, to be turned on and off with a simple voice command.

But as long as we're talking about hypersurveillance and real-time data mining from a personal point of view, how about a television that watches *you* and then transmits data about your viewing habits directly to advertisers and programming executives? Once again, although it may sound far-fetched, the technology exists, and televisions incorporating this feature are *already on the market*, pioneered by manufacturers Samsung and Lenovo. Rolled out at the annual Consumer Electronics Show in Las Vegas in January 2012, these flat-screen televisions can "recognize you and others in the room, automatically logging you into Facebook and pulling up your favorite channels or websites." Michael Learmonth continues:

> Lenovo's TV lets you use the camera as an ID service that blocks access to certain content or channels if a child is in the room. For Samsung's 7500 and 8000 series TVs, all you have to do is say "Hi, TV," when you walk into a room for the TV to turn on and know who's there. . . .
>
> Many people in the living room are multitasking with

other devices. "We're paying for that," said Rex Harris, innovations supervisor at SMGX, a unit of ad agency holding company Publicis Groupe. "If you're looking at other screens, then you're not paying attention. We would like to know if we're getting accurate impressions." Consumers stand to gain too, according to Harris. "The idea is, if the ad is more targeted to you, you will get more value out of it," he said. "When your device knows where you are and knows what you like, it will be a more valuable experience for you."

Indeed, Learmonth is being very circumspect on this issue, and Harris is engaging in outright sophistry when he suggests that the idea is to provide "a more valuable experience" for consumers. This technology is simply ultra-targeted advertising, and the only entity that has anything to gain is the advertiser. Further, both the Samsung and Lenovo televisions have an astonishingly prominent Orwellian presence: a camera that quite clearly and without subterfuge is recording everything happening, using facial recognition technology to identify viewers and scanning the room "to read facial expressions and determine whether you're entertained or bored" (Learmonth).

This isn't technology that's in the experimental stage; it's available right now. And astonishing as it may seem, people are lining up in droves to embrace a television that watches them. Is this really any different from the "telescreens" that monitored social behavior in Orwell's 1984? It may have taken a bit longer to get there—twenty-eight years, to be exact—but it seems we've arrived at the point where many of us want nothing more than to be a part of the web, constantly streaming personal information to central databases and receiving a steady stream of information, advertisements, and product placements in return.

British tech analyst Nick Hide agrees, commenting that, "in

the kind of dystopian insanity that would have George Orwell banging his head on his keyboard, Samsung's newest Smart TVs watch *you* . . . with a built-in camera and mic. The camera can interpret simple gestures—move your hand around to control a cursor and clench your fist to 'click'—and even the different faces of your family members. You can associate permissions with various faces, so if your wee one turns on the telly they'll only be able to watch *CBeebies* [a British children's program]. Voice recognition means you can tell your TV which channel you want to watch and change volume, among other functions," all designed to increase the "advertorial" function of the device.

Kindle and Nook users also get extra attention from their reading devices that they may be unaware of: the portable platforms monitor what they read, how they read it, what portions of the text they highlight, and other patterns of textual consumption. As Alexandra Alter points out, the major online content suppliers know what you're buying and how you use it:

> Barnes & Noble, which accounts for 25% to 30% of the e-book market through its Nook e-reader, has recently started studying customers' digital reading behavior. Data collected from Nooks reveals, for example, how far readers get in particular books, how quickly they read and how readers of particular genres engage with books. Jim Hilt, the company's vice president of e-books, says the company is starting to share their insights with publishers to help them create books that better hold people's attention. . . . [For example,] nonfiction books tend to be read in fits and starts, while novels are generally read straight through, and . . . nonfiction books, particularly long ones, tend to get dropped earlier. Science-fiction, romance and crime-fiction fans often read more books more quickly than readers of literary

fiction do, and finish most of the books they start. Readers of literary fiction quit books more often and tend [to] skip around between books. . . . Pinpointing the moment when readers get bored could also help publishers create splashier digital editions by adding a video, a Web link or other multimedia features. . . . "If we can help authors create even better books than they create today, it's a win for everybody" [Hilt said].

It's certainly a "win" for commercial fiction writers like Scott Turow, who told Alter, "If you can find out that a book is too long and you've got to be more rigorous in cutting, personally I'd love to get the information." Publisher Jonathan Galassi of Farrar, Straus & Giroux vehemently disagreed, telling Alter, "A book . . . can be eccentric, it can be the length it needs to be, and that is something the reader shouldn't have anything to do with. We're not going to shorten *War and Peace* because someone didn't finish it." But in the digital future, reader demand may indeed influence the creation of new texts; it's already forcing popular writers to turn out an increasing amount of material to remain visible in the marketplace, as I've observed elsewhere in this text. Hilt's comments thus seem disingenuous at best: what does he mean by creating "better books"? They may be more commercial, but better?

So Facebook and Google, as well as other social media sites and search engines, aren't the only ones collecting data. In fact, a little-known firm, the Acxiom Corporation, outstrips the more public collectors of consumer statistics. As Natasha Singer wrote on June 17, 2012:

It knows who you are. It knows where you live. It knows what you do. It peers deeper into American life than the F.B.I. or the I.R.S., or those prying digital eyes at Facebook and Google. If you are an American adult, the odds are that it knows things like your age, race, sex, weight, height, marital status, education level, politics, buying habits, household health worries, vacation dreams—and on and on. Right now in Conway, Ark., north of Little Rock, more than 23,000 computer servers are collecting, collating and analyzing consumer data for a company that, unlike Silicon Valley's marquee names, rarely makes headlines. It's called the Acxiom Corporation, and it's the quiet giant of a multibillion-dollar industry known as database marketing. Few consumers have ever heard of Acxiom. But analysts say it has amassed the world's largest commercial database on consumers—and that it wants to know much, much more. Its servers process more than 50 trillion data "transactions" a year. Company executives have said its database contains information about 500 million active consumers worldwide, with about 1,500 data points per person. That includes a majority of adults in the United States. . . . Acxiom's customers have included big banks like Wells Fargo and HSBC, investment services like E*Trade, automakers like Toyota and Ford, department stores like Macy's—just about any major company looking for insight into its customers. For Acxiom, . . . the setup is lucrative. It posted profit of $77.26 million in its latest fiscal year, on sales of $1.13 billion. . . . In essence, it's as if the ore of our data-driven lives were being mined, refined and sold to the highest bidder, usually without our knowledge—by companies that most people rarely even know exist.

Indeed, Acxiom's relative anonymity clearly works to its advantage. It doesn't have much of a public presence, by design; it doesn't directly interface with the consumer, also by design.

Nevertheless, Acxiom is collecting and constantly updating a living database of information on nearly *every* American adult. All this information is neatly packaged according to corporate demand and auctioned off like a virtual mailing list of our wants, needs, and desires—without our even knowing it. And as Singer later found out when she tried to get her own data from Acxiom, it isn't available to individuals; only the companies that want to sell her something, check up on her past, examine her credit, or otherwise intrude on her private life have access.

As Rex Harris noted in his endorsement of surveillance television, if viewers aren't paying attention, advertisers aren't getting their money's worth. It's clear, then, that the real content of twenty-first-century television programming isn't any fictional construct or "reality show"; it's the advertisements and the audience's *response* to those advertisements. The programming is merely something to get viewers to watch the ads. Reality exists only as a database that can be mined for profit. You watch the world through Google glasses, and it watches you, monitoring your every action, word, and emotion; you watch your television, and it watches you. Certainly, this surveillance function will soon be added to iPhones, iPads, and the like; in fact, with Skype, the potential is already there. Why can't anyone hack in on your conversation? Do you really think that Google, Apple, and the other web hyperconglomerates are simply passing your information along from one person to the next without taking inventory? No—*you* are the ultimate streaming program material in the twenty-first century, constantly broadcasting a vast array of information to corporations and advertisers about the way you shop, eat, dress, work, play, and even conduct your relationships with others.

Indeed, the array of streaming data is so vast that Facebook has built a huge "data farm" just sixty miles south of the Arctic

Circle in Lulea, Sweden, to house the vast amount of information it has collected. The site was chosen because the cold weather naturally cools the servers, which have a tendency to overheat and subsequently break down. "There also happens to be a large, fast-flowing river nearby, complete with a large hydroelectric dam that local officials said generates twice as much electricity as the Hoover Dam" (Potter). Well, since it has to put all that information *somewhere*, the Arctic Circle seems to be a congenial location for the billions of megabytes of data that Facebook collects, stores, and puts to commercial use.

And there's a need for such a facility, for the web is expanding at a blinding speed. As Rosa Golijan points out, here's what happens on the Internet *every sixty seconds*:

> Over 6,600 photos will be uploaded to Flickr; about 70 new domains will be registered; over 1,200 new ads will be created on Craigslist; the search engine Google serves more than 694,445 queries; 600 videos are uploaded to YouTube, amounting to 25+ hours of content; 695,000 status updates, 79,364 wall posts and 510,040 comments are published on social networking site Facebook; . . . 168,000,000+ e-mails are sent; 320 new accounts and 98,000 tweets are generated on social networking site Twitter; iPhone applications are downloaded more than 13,000 times; 20,000 new posts are published on micro-blogging platform Tumblr; [the] web browser FireFox is downloaded more than 1,700 times; [the] blogging platform WordPress is downloaded more than 50 times; WordPress Plugins are downloaded more than 125 times; 100 accounts are created on professional networking site LinkedIn; 40 new questions are asked on YahooAnswers.com; 100+ questions are asked on Answers.com; 1 new article is published on Associated Content, the world's

largest source of community-created content; 1 new definition is added on UrbanDictionary.com; 370,000+ minutes of voice calls [are] done by Skype users; 13,000+ hours of music streaming is done by personalized Internet radio provider Pandora; [and] 1,600+ reads are made on Scribd, the largest social reading publishing company.

And that's every *sixty seconds*. Multiply this by the minutes in every hour and the hours in every day, and you can see that the web is exploding with new material. And since these statistics are from a sampling taken on June 16, 2011, one can only assume that the numbers have grown exponentially.

Streaming video downloads form a significant chunk of that traffic. Indeed, according to Dan Hope, as of May 2011, Netflix streaming video alone accounted for an astounding 22 percent of all Internet traffic on *average*; it was even more (30 percent) during prime viewing hours. As Hope notes, a study by Sandvine, a web monitoring service, found that Netflix is responsible for the largest portion of Internet traffic in North America. In fact, "Netflix streaming produces the same amount of traffic as normal web browsing and YouTube videos combined. . . . Netflix movie and TV show streaming accounts for 22 percent of all North American Internet traffic, while standard web page browsing accounts for just under 17 percent and YouTube videos make up 8 percent." These are the latest statistics available, but Sandvine's CEO Dave Caputo predicted that by May 2012, streaming entertainment, "which includes Netflix and other media-streaming services, will make up 55 percent of the peak Internet traffic in North America" (as quoted in Hope). That's a staggering number and indicates a complete change in viewing habits.

Televisions are now routinely connected to Netflix, Hulu, Facebook, and other streaming content providers, so the family flat screen has become the portal for an almost infinite number and a bewildering variety of programming choices. The big three networks still control most prime-time programming, but the audience is skewing toward older members each year. For the eighteen to thirty-four crowd, Hulu, Netflix, and Amazon are the one-stop sources for streaming entertainment, some of which is available for free or at a modest cost per program. As Melanie Gross points out:

This gives ISPs [Internet service providers] an idea of what they have to look forward to in the future. People are going to want more and more bandwidth as the way they consume entertainment changes. The problem with this whole thing is the ISPs, mostly due to the few people who do torrent on a regular basis, have begun putting in [bandwidth] caps. In the states, even the major ISPs who arguably have lots of bandwidth to spare are debating putting in caps anywhere from 150–250 GB. Even in Canada, where Netflix has been hugely successful since it arrived in September [2011], some of the major DSL [digital subscriber line] ISPs are debating bandwidth caps of 25 GB. If you're streaming video, you could go through that in a day. Netflix has already had to lower the quality of its streaming in Canada to deal with the bandwidth caps. Canadian ISPs say that the current infrastructure just isn't meant to handle the amount of internet traffic people are asking for, and that problem is only going to get worse.

This problem is hardly confined to the United States and Canada, as Robert Andrews pointed out in May 2011:

Latin America and Europe are experiencing the same phenomenon but less so—in Europe, video and audio as a proportion of overall traffic has remained stable over the last three years; traffic there is dominated by Bit Torrent, HTTP and YouTube, in that order, with the UK's leading VOD service from BBC iPlayer making up 6.6 percent in Britain. But the UK, Mexico and Brazil can expect the Netflix effect, too, if reports about its planned international expansion are anything to go by. Just six months after launching its first non-U.S. service, in Canada in September, Netflix already accounts for 13.5 percent of peak-time downstream traffic in that country.

These "traffic jams" on the web are a consequence of the rapid growth of Netflix and other streaming video sources. Even music and Kindle or Nook texts take up more room than one might expect, to say nothing of the megabytes consumed by the family photo album or "home movies" shot on video and then downloaded, edited, and ultimately stored on the family computer. Erika Morphy takes the bandwidth analogy even further:

If the Internet is the equivalent of our solar system, then Netflix would be Jupiter, its largest planet [because] the average Netflix subscriber consumes more than a gigabyte of data per day. . . . Broadcasts of the recent Royal Wedding [of Prince William and Kate Middleton in 2011] were telling, from the content providers' perspective. [Lee Brooks of Sandvine noted,] "We saw a number of TV networks take a plunge into this space, providing online videos with additional coverage. It wasn't just the pure stream of events that you typically find on YouTube. Consumers, for their part, wanted to go to the networks with which they were familiar to watch the wedding—but from the Internet." Netflix is a very "sticky" over-the-top service [noted Dan Geiger, vice president of marketing for BroadHop], and differs from such services as Facebook, Skype, WebEx voice and screen sharing, and Pandora, in that its customers are loyal users and willing to pay for content.

Yes, consumers *are* willing to pay for Netflix, and they *are* willing to sit through commercials to stream Hulu and similar services. This pay-per-usage model has thus far eluded Facebook, whose initial public offering in May 2012 was almost immediately undercut by a precipitous drop in the stock's value in the weeks and months that followed (Cutler). Advertisers aren't convinced that Facebook is delivering the 900 million viewers it claims to have access to; or perhaps, while they're chatting back and forth about their personal affairs, they're not buying anything in the process. Further, in August 2012 it was revealed that Facebook has created "more than 83 million fake accounts" that no person has authorized (Prigg). I can speak from personal experience in this regard: I "have" a Facebook page that contains my information copied from Wikipedia and includes my picture, but I never authorized this, don't want it, and would never in a million years subscribe to Facebook, for all the reasons I've outlined throughout this text. Indeed, I regard this as an invasion of privacy that amounts to an implicit endorsement of Facebook, even though I have absolutely no connection to "my" Facebook page and would much prefer that it be taken down. Unfortunately, the only way to do so would be to open an account with Facebook and then log in, which I absolutely refuse to do. So, as far as I'm concerned, "my" Facebook page is entirely fraudulent, but there's no way to remove it, so I just ignore it. No doubt, Facebook created it in the hope that I would see it, feel flattered, and become a member. This, however, will never happen. In the meantime, as of August 2012, Facebook's value had dropped to roughly half of its initial public offering (IPO) valuation: on August 2, 2012, one could buy a share of Facebook for $19.82, whereas the IPO price was $38 per share (Prigg).

Though Facebook is essentially a "data mining" site, transforming that information into dollars is turning out to be more difficult than many supposed. It's a "social" website, and users seem to want to keep it that way. Indeed, many Facebook users are tuning out after a few random postings, opting to be entertained by Netflix and similar content providers rather than sending out messages to a host of phantom "friends." Further, new social media sites such as Care2, Touch, FamilyLeaf, Everyme, and other domains are popping up daily, offering a less intrusive and more manageable alternative to Facebook's draconian regime (Foley, "A Network"). And other forms of entertainment are also competing for viewers' eyes.

YouTube recently launched 100 new channels of programming, all aimed at specific segments of the viewing public and covering every conceivable aspect of society and popular culture. Here's a sampling: *123UnoDosTres*, a "Latino channel offering A-List celebs, music/dance, fashion, sports and telenovelas to a younger bi-lingual audience"; *Alchemy Networks*, an "urban lifestyle network focusing on music, fashion, and storytelling, with [the] biggest brands and celebrities in urban entertainment"; *American Hipster*, "a comedic channel exploring pop culture through the lens of hipsterism"; *Awesomeness*, "from Brian Robbins the producer of *All That!* and *Smallville*"; *Look TV*, "for people who love the excitement and drama of the worlds of fashion, beauty and style, with the creators of *Project Runway*"; *Network A*, an "action sports network dedicated to the athletes who make it happen"; *Black Box TV*, "a scripted series from Anthony E. Zuiker, creator of *CSI*"; *CafeMom Studios*, "a lifestyle network for today's moms, with personal stories, parenting advice, and news, featuring Andrew Shue & Suzanne Somers"; *KickTV*, "connecting soccer fans to the game like never before with daily news, info, entertainment from competitions the world over"; *TBD*, "a place for animal lovers that celebrates the bond between people and their pets, and shows how they enhance our lives"; *ClevverTeVe*, "a daily Spanish language celebrity and entertainment news [channel with] the latest Hollywood headlines [for] a Latin American audience"; *DeepSkyVideos*, offering "a fresh look at the strange and unimaginable depths of space, galaxies, nebulae and other objects"; *Everyday Health TV*, "a 21st century health, fitness and wellness network, where leading experts give advice"; the *Lionsgate Fitness Channel*, "the premier destination for high-quality fitness programming featuring top brands/personalities, and cutting-edge workouts"; *Little Black Dress*, offering "fashion magazine expertise . . . that is entertaining, informative and interactive"; *New Nation Networks*, "a multicultural channel reflecting America's new generation—a hip audience that enjoys comedy, music, and strong points of view"; *Reuters.com*, "a distinctive lineup of talent-driven video programs that reflect global leadership in news, analysis and programming innovation"; *Slate News Channel*, a "topical video [channel that] features [the] irreverent, populist sensibilities of *Slate*"; *Stan Lee's World of Heroes*, featuring "amazing tales about unique characters and extraordinary individuals"; *Start*, a "new video games channel, . . . for anyone who loves or plays games"; the *Comedy Shaq Network* with Shaquille O'Neal, "an urban comedy network featuring stand up comedy, sketch comedy, sitcoms and animation"; the *Wall Street Journal*, "a lifestyle channel . . . covering design, fashion, travel, wine, food and tech"; *Thrash Lab* from Ashton Kutcher, delivering "the best artists in the world [creating] insightful, narrative rich stories"; *u look haute!*, "a channel dedicated to all things related to beauty, style and fashion for teenagers and young adults"; and *YOMYOMF* (You Offend Me, You Offend My Family), which aims to "redefine and give a new voice to the Asian American experience through a unique point-of-view" (all quotes from "YouTube: Channel List"). YouTube doesn't call these programming streams "channels" for nothing—they're the next generation of television, and just like television, they'll be sponsored by advertisers hoping to sell you their products. Indeed, the televisual landscape offered by these various servers, all of which you can subscribe to right now, is so highly commercial that it entirely blurs the line (if one ever existed) between entertainment or information, on the one hand, and out-and-out advertising and product placement, on the other. These channels are really sales platforms, not entertainment or news conduits; they target a specific segment of the viewing audience and then sell, sell, sell.

Don't underestimate these channels; they're pulling in some serious numbers. *YOMYOMF* had 1,319,269 viewers in the week of May 31–June 6, 2012, rising to the number-nine slot in the *Vid Stats X* ratings (the web equivalent of television's Nielsen ratings), as compiled by *Deadline Hollywood*, the online show business authority. Even more impressive is the fact that these are *prelaunch* numbers; *YOMYOMF* didn't officially go online until June 12, 2012, thus demonstrating a hunger for Asian American programming that is not being satisfied by conventional media. *Sourcefed*, a pop media interview channel with comedic overtones, has even more impressive stats; launched in January 2012, it had 8,117,553 viewers during the same May 31–June 6 period, for an astounding total of 112,753,957 views since its debut. Other big winners are *Shut Up Cartoons* (the channel's title is self-explanatory), with 4,258,418 views for the week and a total of 19,052,890 views since its April 2012 launch; *WWE Fan Nation*, a channel devoted to wrestling, with 3,453,807 views for the week and an amazing total of 308,529,212 views since February 2008; the testosterone-fueled *Redbull* channel, with 2,491,461 views for the week and a total of 292,100,924 views since starting up in July 2007; the *Motor Trend* channel, with 2,224,587 views for the week and a total of 178,225,112 lifetime views since September 2008; the satirical channel *The Onion*, with 527,484 weekly views and 124,245,293 lifetime views since January 2008; and the music video channel *The Warner Sound*, with 3,948,297 views for the week and a total of 9,438,832 views since March 2012 (Lieberman, "YOMYOMF Network Leaps into Top 10").

So people are watching, subscribing, and coming back for more; this is just the beginning of an entertainment revolution. Imagine that this was 1950 and these were the ratings for the fledgling television networks. You'd be impressed, and more, you'd be convinced that this mode of entertainment had a future. At an investment of a mere $100 million for 104 channels, all of which contain advertising, this is a smart move for YouTube, even if it signals the end of the "Wild West" era on the web, when content was all viewer produced, not created by corporate sponsors. But that's the direction the web is moving in these days: monetization of content. And with viewer numbers like these, it seems to be working. Very rare is the "homemade" video that cracks the million-viewer mark, to say nothing of millions of *repeat* viewers, week after week. Viewers seem to prefer professionally produced content over home-brew video posting, especially when it's all slotted into thematically disparate channels. Is YouTube TV a success? Indeed it is—a resounding one.

This makes the early days of television in the 1950s and 1960s seem almost halcyon by comparison. Back then, for the most part, you got live programming, old movies and cartoons, with only a few commercials interspersed between. But now sites such as *Daily Motion* won't even let viewers watch a seven-minute cartoon from the 1930s without interrupting it with an advertisement, and they've discovered that people will simply sit there and wait for the commercial to end, eyes glued to the screen. Occasionally, as with coverage of the wedding of Prince William and Kate Middleton—a prepackaged spectacle if ever there was one—all the "content providers" have to do is show up and aim the cameras at the numerous participants, with no staffing required. The "live feed" can be offered without interruption or with "limited commercials," as was the case with the twenty-four-hour streaming video of the Le Mans race on June 11–12, 2011, carried on *Speed.com*. But more often, the ads *are* the content, and many viewers seem absolutely satisfied with that. But how can we possibly sort through this onslaught of material? As Eric Schumacher-Rasmussen asks, "With so much great online video streamed to our PCs, laptops, iPhones, iPads, Android devices, and even our televisions via the Internet, how do we know where the good stuff is? . . . Bruce Springsteen once sang about '57 Channels (And Nothin' On)'; these days, it's thousands of channels and everything is on. Every time I check my Roku, there are channels I didn't even know existed but that I absolutely must watch, in addition to channels I never knew existed and still don't much care about."

Some observers are optimistic about the pervasive influence of Facebook, Netflix, Hulu, YouTube, and other social and entertainment platforms. Gwen O'Keefe, for example, claims that "the digital landscape is a positive place for kids." O'Keefe, "lead author of the American Academy of Pediatrics 2011 report on the impact of social media on children, adolescents, and families, [argues that] 'it promotes a lot of healthy habits like socialization and a sense of connectedness to the greater world and to causes'" (as quoted in Listfield, 9). At the same time, however, she notes that the web is no place for children to explore without some parental guidelines. As Emily Listfield notes, "Fifty-one percent of American teens log on to a social network site more than once a day, and 22 percent log on more than 10 times a day, according to a recent poll by Common Sense Media. You have to be 13 to join Facebook, but children should learn before then not to share personal information" (10).

As of this writing, Facebook is trying to get the preteen audience on board as well, with additional "safeguards" in place, but sharing personal information is what Facebook is all about. Facebook wants everyone to join and divulge all their personal information, which then becomes the permanent intellectual and real property of Facebook. Join when you're seven, and Facebook will know when you turn fourteen or eighteen and target you accordingly. Getting children to sign up for Facebook is the old "hook 'em young" strategy taken to a new extreme; Facebook's "Timeline" feature then chronicles their entire virtual, and real, existence. Facebook is also extremely peremptory in the way it deals with its clients; as just one example, the site converted all the existing e-mail accounts of its members to Facebook e-mail accounts on June 25, 2012, without asking permission beforehand (Constine).

But Facebook is only one aspect of the online world; in addition to the various streaming media sites, social networks, and information sources such as Wikipedia, there's the gaming community, which has taken to the web with a vengeance. Video games were once confined to arcades; now, everyone can play on their laptops, iPhones, iPads, or other mobile devices. As Listfield notes:

When kids play video games, that little pleasure chemical dopamine also kicks in. The intermittent reinforcement that games provide—you win a little, you want to play more—is similar to gambling, and for some kids, just as addictive. Ninety-two percent of kids ages 8 to 18 play video games, and 8.5 percent can be classified as addicted, meaning their play interferes with the rest of their lives. According to Douglas Gentile of Iowa State University, lead author of a 2011 study on video game addiction, 12 percent of boys and 3 percent of girls who play will get addicted. . . . Increased game play is related to poorer school performance as well as higher rates of obesity. . . . For every hour children are spending on games, they are not doing homework, exercising, or exploring. (14)

There are other consequences as well:

A 2010 study by the Kaiser Family Foundation found that students 8 to 18 spend more than 7.5 hours a day engaged with computers, cell phones, TV, music, or video games. Forty percent of kids in middle school and high school say that when they're on the computer, most of the time they're also plugged into other media. The effects this multitasking has on still-forming brains can be positive and negative. "The prefrontal cortex, which is essential for social behavior, planning, reasoning, and impulse control, is not fully developed until the early 20s," says Jordan Grafman of the Kessler Foundation Research Center. "Its development is largely dependent on what activities you do." (Listfield, 19)

The long-term effects of this inundation of images and "advertorial" information become clear. Further, it can lead to a syndrome of instant narcissism: everyone is an instant "star" on Facebook, the center of their own universe, with a host of online friends who "like" them—until the criticism inevitably starts. Go to any content website that allows comments, and you will find a string of brief, often illiterate notations by a host of users, often hiding behind their screen names, firing insults, racist and sexist epithets, and sheer venom at the person who posted the initial article. Positive comments are rare, but they can lead users to believe they're omniscient "experts" in fields they really have little or no understanding of at all. As British novelist Ian McEwan noted in a 2007 interview, "I don't have much time for the kind of [Internet] site where readers do all the reviewing. Reviewing takes expertise, wisdom and judgment. I am not fond of the notion that anyone's view is as good as anyone else's." But on the web, that's exactly what's happening; everyone is an instant critic, whether they have the credentials or not.

Indeed, in July 2000 the Sony Corporation created a fictitious critic, "David Manning," specifically to offer rave reviews of Sony/Columbia films, such as Brian Helgeland's *A Knight's Tale* and Luke Greenfield's comedy *The Animal* (both 2001), which real critics had panned. John Horn reported the deception in the June 2001 issue of *Newsweek*. In August 2005 Sony agreed to a financial settlement of $1.5 million for initiating the hoax. Although the creation of "Manning" was obviously unscrupulous, given the current level of critical discourse, such a scheme isn't all that surprising (Elsworth). How reliable are many of the "real" reviews by the utterly uninformed roaming the Internet? Why should anyone believe them? The creation of "David Manning" was, in many ways, a portent of the future.

While McEwan derides what many observers call "fanboy" culture, the question arises: who, exactly, is out there watching all the content that streaming, in all its forms, so effortlessly provides? Most people assume that the web audience is overwhelmingly male and adolescent, or at least male and between the ages of eighteen and thirty-five. Wrong. As researcher Genevieve Bell noted on May 10, 2012, as part of the "Big Ideas" series of lectures on Australia's Radio National:

It turns out women are our new lead adopters. When you look at Internet usage, it turns out women in Western countries use the Internet 17 percent more every month than their male counterparts. Women are more likely to be using the mobile phones they own, they spend more time talking on them, they spend more time using location-based services. But they also spend more time sending text messages. Women are the fastest growing and largest users on Skype, and that's mostly younger women. Women are the fastest [growing] category and biggest users on every social networking site with the exception of LinkedIn. Women are the vast majority owners of all Internet enabled devices—readers, healthcare devices, GPS—that whole bundle of technology is mostly owned by women. (As quoted in Madrigal)

You certainly wouldn't know this from attending ComicCon or any of the other ever-proliferating host of fan conventions, but it's true: women are the primary movers and shapers of web culture, even if advertisers are slow to recognize that fact. As Bell points out, "If you want to find out what the future looks like, you should be asking women. And just before you think that means you should be asking 18-year-old women, it actually turns out the majority of technology users are women in their 40s, 50s and 60s . . . [and] those turn out to be the heaviest users of the most successful and most popular technologies on the planet as we speak" (as quoted in Madrigal).

And this gender difference drives *actual* consumer consumption, as opposed to perceived consumption. That is why YouTube's "movies on demand" service is skewed heavily toward romantic comedies ("rom coms"), why romance novels are such a pervasive segment of the Nook and Kindle streaming landscape, and why the Gem Shopping Network, with its endless, hypnotic turntables of jewelry accompanied by nonstop offscreen patter from its hard-sell salesmen, is so popular. Likewise, this accounts for the appeal of "reality" television shows like *Bridezillas* and the immense popularity of such disparate media icons as Ellen De-Generes, Kim Kardashian, Lady Gaga, and, even after all these years, Madonna.

Programming costs money, but eventually, Apple or some other megaconcern might just buy up content providers outright, eliminating the need for continual negotiation. Just as NBC wound up buying Universal Studios when the cost of renewing the then-profitable *Law and Order* franchises outstripped the actual value of the studio itself, Apple could simply *buy* Hollywood and take control of the content providers in a definitive fashion. Apple has roughly $100 *billion* in its cash reserve fund, and as Erick Schonfeld notes somewhat dreamily:

Apple wants to bring Hollywood into people's homes in an entirely new way. Hence all the chatter lately of a real Apple TV in the works. However, before TVs can become more than a hobby for Apple, there is a major roadblock it must get past. The reluctance of Hollywood to license its best movies and TV shows at the price Apple wants to pay. In that light, all the cash Apple has been hoarding and building up for years now becomes more intriguing. Its staggering piles of money now total $97.6 billion to be precise. What are they going to do with all that cash? One thing they could do is buy their way into Hollywood. Think about it for a second. Today, Apple could literally buy Time Warner ($38 billion market cap), Viacom ($29 billion), and Dream-works ($1.6 billion) combined, and still have $30 billion left over. If it waits a few more quarters it could snap up News Corp ($49 billion) as well. Only Disney, which is worth $70 billion, would take a while longer to save up for.

Another factor is that anonymity—as we once knew it—is a thing of the past. As Brian Stelter notes in "Upending Anonymity":

Not too long ago, theorists fretted that the Internet was a place where anonymity thrived. Now, it seems, it is the place where anonymity dies. The collective intelligence of the Internet's two billion users, and the digital fingerprints that so many users leave on web sites, combine to make it more and more likely that every embarrassing video, every intimate photo, and every indelicate e-mail is attributed to its source, whether that source wants it to be or not. This intelligence makes the public sphere more public than ever before and sometimes forces personal lives into public view.

Further, a web app appropriately named Creepy—yes, that's right, Creepy—allows anyone to track your every movement on the web, using information you've posted on Flickr, Twitter, and other social media sites. As Rosa Golijan discovered:

Creepy [allows users] to pinpoint anyone's location, using geographic data embedded within shared photos. . . . All you have to do is type in a person's Twitter or Flickr username, and hit the "Geolocate Target" button. The app will then gather all the geographic information available online, via photos that the "target" has shared online. . . . Whenever someone shares a photo taken with his or her smartphone, services like Flickr, Yfrog and Twitpic automatically record the location where the shot was taken, and store that geo-tag in the image's EXIF data. Creepy pulls up that data and places it onto maps. . . . [There's] also a website called *I Can Stalk You* which demonstrat[es] how geotagged photos —if combined with details gleaned from tweets—can reveal information [such as where you live, who else lives there, your commuting patterns, where you go for lunch each day, who you go to lunch with, and why you and your attractive coworker like to visit a certain nice restaurant on a regular basis] which could easily be abused by someone with nefarious intents.

In short, it's a big, wide, wonderful world out there, but you have to be careful about sharing your information with *too* many people. As Yiannis Kakaras, inventor of the Creepy app, notes, why are you "publishing [this information] in the first place?" (as quoted in Golijan, "'Creepy' App"). You not only have to be careful about privacy settings on Facebook and other social media sites; just the act of walking around with a cell phone (with automatic GPS tracking) indicates your immediate location. If you're not at home and you're not in your new car—because, according to GPS, you're moving so slowly that you're obviously on foot—then your car could very well be in your driveway. Didn't you post a picture of it just the other day? Isn't it a Porsche? Don't you live at 2400 South Central Drive? Why don't I stop over there right now and see if it's unlocked?

There's also a new iPhone app called Scene Tap that scans your local bar to see if the crowd is "your kind of people." Launched in twenty-five bars in the San Francisco area, Scene Tap "scan[s] the faces of patrons . . . across the city to determine their ages and genders. Would-be customers then can check their smartphones for realtime updates on the crowd size, average age, and men-to-women mix" (Wohlsen). In many ways, Scene Tap is the logical extension of a similar device described in Gary Shteyngart's comic novel *Super Sad True Love Story* in 2010. In her review of the novel, Michiko Kakutani describes Shteyngart's vision of the future: "Everyone carries around a device called an äppärät, which can live-stream its owner's thoughts and conversations, and broadcast their 'hotness' quotient to others." It's just another example of how rapidly technology is outstripping the so-called bounds of reality. Then, such a device was fiction; now, it's fact. Scene Tap is aided by the fact that the iPhone's built-in camera app already "highlight[s] a person's face on the screen with a green box before the photo is snapped[,] and Apple's iPhoto software will try to recognize the faces of the people in users' pictures to categorize photos automatically by who's in each shot" (Wohlsen). As Lee Tien, an attorney with the Electronic Frontier Foundation, notes, "Ten years ago, if I walked down the street and took a picture of someone I didn't know, there was little I could do to find out who that person was. Today, it's a different story" (as quoted in Wohlsen). And really, one doesn't need to *do* anything. Facial recognition software will do the job for you, in milliseconds.

There are, of course, many positive aspects to Facebook, cell phone video cameras, and other instantaneous communication and messaging devices, as demonstrated during the so-called Arab Spring, when conventional news media couldn't gain access to the various countries in question. Cell phone videos brought home the urgency of the situation to hundreds of millions of concerned viewers around the world. In Pakistan, Libya, Egypt, Iran, Iraq, and elsewhere, cell phones, Facebook, and Twitter can be invaluable sources of news and information, not only for the citizens of those countries but also for news-gathering organizations. Unfortunately, those same images and tweets can be used by repressive regimes to retaliate against those who oppose them, utilizing facial recognition software to identify them and punish them on an individual or mass basis.

Responsible use of social media is one thing, but when it comes to relying on it constantly, around the clock, a rather bleak pattern begins to emerge. The Chennai Social Media group concluded in a 2011 survey that excessive—that is, constant and un-relenting—use of Facebook, Twitter, LinkedIn, FourSquare, and other social media sites can have serious consequences. As Chennai's "Negative Aspects of Social Networking" notes:

> Social networking has become inevitable in modern life, with many people using it to stay in touch and companies using [it] to market themselves. However, the negative aspects of social media are [many, including] loss of productivity (individuals who spend a lot of time on social networking lose valuable time that can be devoted [to] doing something creative and useful); lack of prioritizing (checking others' comments, messages and statuses can result in not focusing on one's own task, which can lead to poor performance in [the] office); diminishing social skills (communicating solely with virtual friends can result in people not finding enough time to socialize with people around them); [and] social media overkill (social media, being viral and possessing the potential of reaching out to millions, instantly carries the risk of being abused. Any negative information or content about anyone or anything can be circulated [instantaneously] through social networking platforms).

To keep up with the masses of data being generated, it's no longer enough for professionals, or even casual or recreational users of the web, to have one screen; in the era of the "data deluge," multiple screens are becoming increasingly common and almost a necessity. Having more than one monitor at your disposal makes the work go faster. Certainly, this is nothing new. In the 1970s, as a video editor for the Los Angeles production company TVTV, I would regularly "pyramid" three television monitors with the video dailies from the day's multicamera shoot and, with the aid of an assistant, view them at ten times their normal frame rate, identifying usable shots as they flew by, giving me a sense of what portions to assemble for a rough cut of the material. It was a common practice even then, when masses of material had to be rapidly edited on a daily basis.

But now we're talking about text, numbers, images, hyperlinks, and tweets—an enormous amount of disparate material to organize—not just a stream of images. As Matt Richtel reports, "Workers in the digital era can feel at times as if they are playing a video game, battling the barrage of e-mails and instant messages, juggling documents, web sites and online calendars. To cope, people [are adding] a second computer screen. Or a third. This proliferation of displays is the latest workplace upgrade, and it is responsible for the new look at companies and home offices—they are starting to resemble mission control" ("In Data Deluge"). And in a world of constantly streaming information, the speed with which we can absorb that information is a crucial factor. Our personal perception is being ratcheted up to move at mechanistic rates; how else can we keep up with the onslaught of data cascading into our virtual world?

Even the "address book" for the web—the suffixes .org, .net, and so on—is running out of space. The Internet Corporation for Assigned Numbers and Nanos (ICANN), which assigns new web addresses to corporations, countries, and individuals worldwide, revealed on June 13, 2012, that it has reviewed no fewer than 1,930 applications for new web suffixes, including ".blog, .book, .cloud, .docs, .phd, and .wow." The new domain suffixes "will take their place alongside the original [ones] (.com, .edu, .net, .org, and .mil); subsequent [suffixes] created over a decade ago (.aero, .biz, .coop, .info, .int, .museum, .name, .pro); and recent additions (.asia, .cat, .jobs, .mobi, .tel, .travel, and .xxx) as legal entities in the domain name system." Further, "66 of [the applications] are for geographic names, and [many more] are in non-Latin alphabets, such as Arabic, Chinese, and Cyrillic. This is the first time that non-Latin [suffixes] have been evaluated for implementation in the domain name system, an event that underscores the increasing internationalization of the internet" (Claburn). Amazon has even applied for an .lol suffix (short for "laugh out loud," as everyone surely knows) in the latest Internet land rush, which, as Thomas Claburn points out, is truly international in scale. And there's going to be a need for all those new addresses, because new sources of programming, entertainment, infotainment, and advertising keep popping up, proliferating in a never-ending succession of new, eager-to-please websites.

Streaming video games continue to proliferate and mutate into subvariations on the web, with such new arrivals as *Beyond: Two Souls, Call of Duty: Black Ops II, Halo 4, The Last of Us, Simcity, Watch Dogs,* and *Xcom: Enemy Unknown*—many of them sequels to or updates of earlier versions—vying for the public's attention (Schiesel). The *TMZ*-style gossip industry continues to churn out cheap broadcast pods for mass entertainment (Rutenberg, 1), and hundreds of thousands of unaffiliated videos on YouTube and other streaming websites, produced with varying degrees of professionalism or amateur enthusiasm, relentlessly seek new viewers, more exposure, and greater visibility—anything to bring them to the center of the streaming vortex. It's all part of the cyclone of images, text, and information that we call the web, and it keeps expanding every day.

As Joichi Ito, director of the MIT Media Lab, puts it, "The Internet isn't really a technology. It's a belief system, a philosophy about the effectiveness of decentralized, bottom-up innovation. And it's a philosophy that has begun to change how we think of creativity itself." Nothing stands still. Everything is constantly being revised, restructured, rebooted, reconfigured, redesigned, and reimagined. Today's frontiers of thought will be tomorrow's backwaters, like a city rapidly expanding into the suburbs and then beyond that into the country and then beyond that, until all the various "cities" find themselves inextricably intertwined. Predictions are risky by nature, but one thing is certain: the streaming web will continue to expand indefinitely, and the connections to our daily life will become even more pronounced.

Cultural critic Claire Suddath predicts that in 2050 India will be the most populous nation, with China in second place and the United States third. In Africa, according to the U.S. Census Bureau, Nigeria and Ethiopia will experience the greatest per capita population increase: from 166 million people to 402 million for Nigeria, and from 91 million to 278 million for Ethiopia. Life in Russia will be less promising, with its population dropping from 139 million to just 109 million by 2050. But overall, the world's total population will continue to rise, reaching an estimated 9.4 billion human beings by the middle of the twenty-first century. And we're all going to be interconnected by commerce, particularly when it comes to the movies: the AMC Theater chain in the United States was sold in May 2012 to the Chinese Wanda Corporation for $2.6 million. The plan is for some "Hollywood" films to be made in China, with American actors and Chinese crews, now that they have a guaranteed theatrical release through AMC; in addition, Wanda hopes to bring more native Chinese films to the United States, and the AMC acquisition will certainly help in that regard ("China's Wanda").

So filmmaking, and visual image production generally, has now crossed all artificial national boundaries, and citizens of the future will be interconnected in ways we simply can't begin to imagine right now, using instrumentalities that make even our most sophisticated information storage and retrieval methodologies seem utterly archaic by comparison. Holographic plays, movies, and concerts will project performers anywhere in the world with perfect, three-dimensional physicality; Real 3-D for motion pictures, as it's called, will become as distant in our collective memory as 1950s Natural Vision 3-D; we'll do away with the screen altogether. Even though handheld devices will dominate our culture, giving each person access to a vast, unimaginable treasure house of entertainment and information, public gatherings—where people can share an experience in the most direct sense—will still be a vital part of the culture.

As technology develops, we might move beyond voice commands to the power of thought itself—thought transformed into action through electrical impulses alone. DVDs, CDs, conventional motion picture film, vinyl records, and chemical photography will be entirely forgotten, except by the archaeologists of twentieth-century culture. Everything will be online all the time, streaming constantly through our television sets, cell phones, iPads, and whatever new devices will be created in the twenty-first-century technological revolution. Vast corporations or mega-conglomerates that gobble up every morsel of content in the hope of reselling it at a profit will control much of this information.

Indeed, a recent U.S. Supreme Court decision seems to lean in that direction. In a 6–2 ruling, the Court found that "Congress may take books, musical compositions, and other works out of the Public Domain, where they can be freely used and adapted, and grant them copyright status once again" (Kravets). Once, when a work entered the public domain, it remained there forever; now, it seems that certain works can be "re-copyrighted" in the United States. As David Kravets wrote of those justices who disagreed with the decision:

In dissent, Justices Stephen Breyer and Samuel Alito said the legislation goes against the theory of copyright and "does not encourage anyone to produce a single new work." Copyright, they noted, was part of the Constitution to promote the arts and sciences. The legislation, Breyer wrote, "bestows monetary rewards only on owners of old works in the American Public Domain. At the same time, the statute inhibits the dissemination of those works, foreign works published abroad after 1923, of which there are many millions, including films, works of art, innumerable photographs, and, of course, books—books that (in the absence of the statute) would assume their rightful places in computer-accessible databases, spreading knowledge throughout the world."

As always, information is prized, fought over, contested, and sought by those who would attempt to "monetize" every aspect of the streaming world. But despite the hegemonic efforts of the ruling corporate masters of the web, as with all creative enterprises, true innovation comes from the margins, and it is individuals who will continue to drive innovation in streaming culture. With Wikipedia, the Internet Archive, and numerous other databases of films, videos, texts, and music proliferating on the web in almost bewildering profusion, we can be certain of one thing: No one person can ever control it. No one corporation can ever monopolize it because, as Yeats would put it, "the center cannot hold," because there *is* no center. The world of streaming is expanding second by second, image by image, one idea after another.

And yet, as this book goes to press, there is a new and even more ominous development in the wings: the devaluation of the liberal arts in favor of sheer technological expertise and nothing more—or technology *as* content, rather than a means to distribute content. In a deeply disturbing and shortsighted op-ed piece in *USA Today*, tech guru Marc Andreessen told writer Tim

Mullaney that "the spread of computers and the Internet will put jobs in two categories. People who tell computers what to do, and people who are told by computers what to do." In this brave new world, only the techies will survive. For Andreessen, "there's no such thing as median income; there's a curve, and it really matters what side of the curve you're on. There's no such thing as the middle class. It's absolutely vanishing." Further, according to Andreessen, those who study the liberal arts are "out of luck."

As Andreessen told Mullaney, he advises college freshmen to "study STEM (science, technology, engineering and math). . . . In liberal arts, only the best of the best will make top dollar—a person will have to be good enough that his book is a best seller or her song goes global, or he'll have to be smart enough to apply philosophy to corporate strategic thinking." This ignores the entire aspect of art as an essentially noncommercial enterprise, designed to push the boundaries of thought and perception in ways that the dominant society must inevitably resist. Not everyone aspires to write a best seller or to have a top-ten pop song; many work on poems, paintings, sculptures, videos, and other forms of art with no thought of financial gain. But for those who dominate tech in the twenty-first century, the bottom line is always paramount. One can only hope that not everyone shares this worldview, and in fact, there's good reason to think that many observers reject this model, for streaming technology puts the arts within reach of everyone, and not everyone is obsessed with making money as the ultimate goal.

There is no beginning to the streaming universe, and no end. Anyone can upload anything, and despite attempts to control it, the web is something like a blob of mercury, always eluding any fixed identity. Those who are now first may one day be last, and those who have yet to participate in the web may one day come to redefine it entirely. The vast storehouse of information that we possess as citizens of the earth has at last been uploaded for everyone on the planet to use, legally or illegally. The web gives a voice to those who were once silenced by the conventional media and utterly reshapes the landscape of books, movies, music, and social discourse. What we have seen is only the beginning. The web now streams out to the world, offering information, entertainment, and enlightenment in countless languages and through myriad cultures. And it brings the world to us, whether mediated by the forces of commercialism or altruistic entrepreneurship, by the purest artistic impulse or the most mercantile business models. We can be certain of only one thing as we move toward the future: it's a streaming world, and there's no turning back. Welcome to the twenty-first century and the new information grid, which is already reshaping the world in its entirety and will continue to do so in ways that we cannot, for the moment, even imagine.

Acknowledgments

First of all, I want to thank Gwendolyn Foster for suggesting that I write this volume, and Anne Dean Watkins for commissioning it. I also want to thank my colleagues Christopher Sharrett, Jon Kraszewski, Marco Abel, Susan Belasco, Michael Downey, Greg Ostroff, Laura White, Stephen Behrendt, Stephen Buhler, David Sterritt, Mikita Brottman, Rebecca Bell-Metereau, Valérie Orlando, Jan-Christopher Horak, Steven Shaviro, Ted Kooser, and many, many others for their input, either directly or through informal discussions on the subject of streaming media.

I also want to thank the editors of *Flow: A Journal of Television and New Media* (particularly Alfred Martin, co–managing editor), published by the Department of Radio, Television, and Film at the University of Texas at Austin, for permission to reprint brief portions of my essays on streaming that originally appeared there, specifically: "Some Notes on Streaming," *Flow* 14.1 (June 9, 2011); "Red Boxes and Cloud Movies," *Flow* 14.4 (July 21, 2011); "How Long Will It Last, and Do You Really Own It?" *Flow* 14.7 (September 3, 2011); "I'm Not Here," *Flow* 15.4 (December 5, 2011); "The Great Wikipedia Blackout, the Stop Online Piracy Act, and You," *Flow* 15.7 (February 27, 2012); and "Film, Nostalgia, and the Digital Divide," *Flow* 15.12 (May 19, 2012).

Some additional material came from the paper "Gently Down the Stream: The New Era of the Moving Image," presented at the 52nd Annual Conference of the Society for Cinema and Media Studies in Boston, Massachusetts, March 22, 2012, and from the article "On the Value of 'Worthless' Endeavor," *College Hill Review* 8 (Summer 2012), reprinted by kind permission of the journal's editor, James Barszcz. I also incorporated some material from my blog, *Frame by Frame*.

I particularly want to thank Dana Miller for her superb job of typing the original manuscript, Christopher DeMarco for his help in compiling the bibliography, Linda Lotz for a great copyediting job, and Jennifer Holan for creating the index.

Works Cited

Adams, Scott. "People Who Don't Need People." *The Scott Adams Blog.* 27 May 2011. Web. 5 Oct. 2011.

Alimurung, Gendy. "Movie Studios Are Forcing Hollywood to Abandon 35mm Film. But the Consequences of Going Digital Are Vast, and Troubling." *LAWeekly.com.* 11 Apr. 2012. Web. 12 Apr. 2012.

Alter, Alexandra. "Your E-Book Is Reading You." *WallStreetJournal. com.* 28 June 2012. Web. 26 July 2012.

Andrews, Robert. "The Netflix Effect: Video Gobbling up Internet Traffic." *Paidcontent.org.* 18 May 2011. Web. 18 May 2011.

AOL, Google, Reddit, Wikipedia, et al. "Letter Opposing H.R. 3261, the Stop Online Piracy Act, 15 Nov. 2011." Web. 26 June 2012.

Baldwin, Howard, and Ed Oswald. "Google's Project Glass Eyewear: Next Big Google Flop or Hit?" *PCWorld.com.* 5 Apr. 2012. Web. 13 June 2012.

Barnes, Brooks. "At Disney, It's Tough to Find a Chief." *NYTimes.com.* 25 Apr. 2012. Web. 29 Apr. 2012.

———. "Web Deals Cheer Hollywood, Despite Drop in Moviegoers." *NYTimes.com.* 24 Feb. 2012. Web. 25 Feb. 2012.

———. "Movie Studios See a Threat in Growth of Redbox." *NYTimes. com.* 6 Sept. 2009. Web. 26 June 2012.

Battaglio, Stephen. "Network versus Technology." *TV Guide.* 18 June–1 July 2012, 7. Print.

Baughman, Brent. "The Art of the Modern Movie Trailer." *NPR.* 15 Jan. 2012. Web. 8 June 2012.

Bennett, Matt. "Live Streaming Video Delivery over Mobile Released by Online Video Platform Company." *Voped Video Platform Solutions.* 9 June 2011. Web. 4 Oct. 2012.

Bensinger, Greg. "Netflix Users' Taste for Movie Streaming Continues to Fall." *WallStreetJournal.com.* 12 July 2012. Web. 26 July 2012.

Bilton, Nick. "With Apple's Siri, a Romance Gone Sour." *NYTimes.com.* 15 July 2012. Web. 26 July 2012.

———. "Indiscreet Photos, Glimpsed Then Gone." *New York Times.* 7 May 2012, B4. Print.

———. "A Rose-Colored View May Come Standard." *NYTimes.com.* 4 Apr. 2012. Web. 13 June 2012.

———. "Behind the Google Goggles, Virtual Reality." *NYTimes.com.* 22 Feb. 2012. Web. 5 Mar. 2012.

Block, Alex Ben. "*Dark Knight Rises* Director Christopher Nolan Isn't a Fan of Digital." *Hollywoodreporter.com.* 9 June 2012. Web. 9 June 2012.

———. "Senate Committee Votes to Make Illegal Streaming of Movies, TV a Felony." *Hollywoodreporter.com.* 16 June 2011. Web. 17 June 2011.

Bosman, Julie. "Writer's Cramp: In the E-Reader Era, a Book a Year Is Slacking." *NYTimes.com.* 12 May 2012. Web. 16 May 2012.

Bowles, Scott. "CinemaCon Bids Farewell to Film." *USA Today.* 30 Apr. 2012, 3D. Print.

Brown, Damon. "Google Glasses Are a Prescription for Disaster." *PCWorld.com.* 22 Feb. 2012. Web. 8 June 2012.

Buñuel, Luis. "Pessimism." In *An Unspeakable Betrayal: Selected Writings of Luis Buñuel.* Trans. Garrett White. Berkeley: University of California Press, 1995: 258–63. Print.

Burns, Matt. "Tim Cook: Apple TV Is Still a Hobby, But I Couldn't Live without It." *TechCrunch.com.* 24 Jan. 2012. Web. 5 Mar. 2012.

"Business Development for Entertainment Technology." *MKPE Consulting LLC.com.* 2012. Web. 5 Mar. 2012.

Calamia, Matthew. "Apple Aims to Stream Movies from iCloud." *Mobiledia.* 1 June 2011. Web. 26 June 2012.

Carmody, Tim. "How and Where We Watch Netflix, Hulu and YouTube." *Wired.com.* 29 July 2011. Web. 26 June 2012.

Carr, Austin. "Newspaper's Digital Apostle." *New York Times.* 14 Nov. 2011, B1, B6. Print.

————. "Is Streaming Video Cannibalizing Amazon's DVD Sales?" *Fastcompany.com*. 6 June 2011. Web. 8 June 2011.

Cashill, Bob. "No Concessions: Digital Disaster at the New York Film Festival." *Pop Dose*. 30 Sept. 2012. Web. 4 Oct. 2012.

"Charlie Sheen Happy about *Anger Management*, with FX Sitcom Poised for Renewal." *Associated Press*. 28 July 2012. Web. 1 Aug. 2012.

"China's Wanda Buying US Cinema Chain AMC for $2.6 Billion." *Reuters*. 21 May 2012. Web. 8 June 2012.

Chmielewski, Dawn C. "YouTube Movie Rental Service Gains 3,000 Titles." *Los Angeles Times*. 10 May 2011, B3. Print.

Chozick, Amy. "Time Inc. to Sell Its Magazines on Apple's Newsstand." *New York Times*. 14 June 2012, B2. Print.

Churner, Leah. "'This Is DCP': The Big Screen Test—New York." *Villagevoice.com*. 28 Feb. 2012. Web. 5 Mar. 2012.

Cieply, Michael. "Many More Indie Films Are Released, But Not Very Widely, Study Finds." *NYTimes.com*. 9 Apr. 2012. Web. 3 June 2012.

————. "The Afterlife Is Expensive for Digital Movies." *NYTimes.com*. 23 Dec. 2007. Web. 5 Mar. 2012.

C. K., Louis. "A Statement from Louis CK." *LouisCK.net*. 15 Dec. 2011. Web. 8 June 2012.

Claburn, Thomas. "New Domain Names Tempt Amazon, Google, Microsoft." *Informationweek.com*. 13 June 2012. Web. 13 June 2012.

Condon, Stephanie. "SOPA, PIPA: What You Need to Know." *CBS News*. 18 Jan. 2012. Web. 4 Oct. 2012.

Considine, Austin. "I'm Not Real, But Neither Are You." *New York Times*. 22 May 2011, ST6. Print.

Constine, Josh. "Facebook Hides Your Email Address Leaving Only @ Facebook.com Visible. Undo This Poppycock Now." *TechCrunch.com*. 25 June 2012. Web. 26 June 2012.

Cox, Tony. "Streaming Video Means Prime Time All the Time." *NPR*. 2 Aug. 2011. Web. 2 June 2012.

Cutler, Kim-Mai. "Zuckerberg on Facebook's IPO: Stock Performance Has Been 'Disappointing.'" *TechCrunch.com*. 11 Sept. 2012. Web. 4 Oct. 2012.

Davidson, Adam. "How Does the Film Industry Actually Make Money?" *NYTimes.com*. 26 June 2012. Web. 26 July 2012.

"Digital Cinema Package." *Wikipedia.org.* 1 Feb. 2012. Web. 5 Mar. 2012.

"Digital Cinema Technology: Frequently Asked Questions." *MKPE. com.* Jan. 2012. Web. 5 Mar. 2012.

"Dish Summer Sales Event." Advertising circular. Summer 2012. Print.

Dixon, Wheeler Winston. "'Let the Sleepers Sleep, and the Haters Hate': An Interview with Dale 'Rage' Resteghini." *Quarterly Review of Film and Video* 29.1 (2012): 1–11. Print.

———. "Film, Nostalgia, and the Digital Divide." *Flow* 15.12 (19 May 2012). Web. 8 June 2012.

———. "Gently Down the Stream: The New Era of the Moving Image." Paper presented at the 52nd Annual Conference of the Society for Cinema and Media Studies, Boston, 22 Mar. 2012.

———. "The Great Wikipedia Blackout, the Stop Online Piracy Act, and You." *Flow* 15.7 (27 Feb. 2012). Web. 5 Mar. 2012.

———. "I'm Not Here." *Flow* 15.4 (5 Dec. 2011). Web. 5 Mar. 2012.

———. "How Long Will It Last, and Do You Really Own It?" *Flow* 14.7 (3 Sept. 2011). Web. 5 Mar. 2012.

———. "Red Boxes and Cloud Movies." *Flow* 14.4 (21 July 2011). Web. 5 Mar. 2012.

———. "Some Notes on Streaming." *Flow* 14.1 (9 June 2011). Web. 5 Mar. 2012.

Dodd, Christopher. "Statement on the So-Called 'Blackout Day' Protesting Anti-Piracy Legislation." MPAA website. 17 Jan. 2012. Web. 5 Oct. 2012.

Dolor, Solon Harmony. "*Newsweek* May Go Digital-Only as Owners Consider Cost-Cutting Decisions." *SocialBarrel.com.* 26 July 2012. Web. 26 July 2012.

Donahue, Deirdre. "E-Book Sales Revenue Officially Tops Revenue from Hardcover." *USAToday.com.* 21 June 2012. Web. 21 June 2012.

Eaton, Kit. "For Starters, These Are a Few of My Favorite Apps." *New York Times.* 14 June 2012, B6. Print.

"Echo Smartpen." *SkyMall.* N.d., 161. Print.

Edwards, Cliff, and Sarah Rabil. "Netflix Surges on Streaming-Only Option, DVD Price Increase." *Bloomberg News.* 22 Nov. 2010. Web. 12 Mar. 2011.

Elsworth, Catherine. "Sony Ordered to Pay $1.5m for Film-Goer Hoax." *Telegraph.* 5 Aug. 2005. Web. 13 Aug. 2012.

Evans, Suzy. "New AFTRA Video Game Contract Includes First Streaming Payment." *Hollywoodreporter.com*. 10 June 2011. Web. 10 June 2011.

Farivar, Cyrus. "Fox, NBC Universal Sue Dish over Ad-Skipping DVR Service." *Arstechnica.com*. 24 May 2012. Web. 12 June 2012.

Finke, Nikki. "Major Hollywood Studios Market Share: 2011." *Deadline.com*. 18 Dec. 2011. Web. 8 June 2012.

Fischer, Pauline. "Temporary Removal of Sony Movies through Starz-Play." *Netflix Blog*. 17 June 2011. Web. 5 Oct. 2012.

Fitzgerald, Drew. "Netflix to Stream Miramax Movies." *WallStreetJournal.com*. 17 May 2011. Web. 21 May 2011.

Fleming, Mike. "Floundering Blockbuster Bought by Dish Network for $320 Million." *Deadline.com*. 6 Apr. 2011. Web. 2 May 2011.

Foley, Nick. "A Network That's All in the Family." *USAToday.com*. 8 Aug. 2012. Web. 13 Aug. 2012.

———. "Teens Turn from Facebook to Fresher Social-Media Sites." *USAToday.com*. 20 June 2012. Web. 21 June 2012.

Franzen, Carl. "Google Glasses Will Be Available for Users with Prescriptions." *TPMideaLab*. 16 Apr. 2012. Web. 13 June 2012.

Friedman, Wayne. "Heavy Streaming Video Viewers Watch Less TV, Nielsen Says." *Mediapost.com*. 14 June 2011. Web. 19 June 2011.

Fritz, Ben, Joe Flint, and Dawn C. Chmielewski. "Starz to End Streaming Deal with Netflix." *LATimes.com*. 2 Sept. 2011. Web. 8 June 2012.

"From Paper to Email in Seconds." *SkyMall*. N.d., 29. Print.

Gardner, Eriq. "Warner Bros. Wins Key Legal Ruling Impacting All *Wizard of Oz* Remakes." *Hollywoodreporter.com*. 6 July 2011. Web. 8 July 2011.

Gayomali, Chris. "Amazon Now Sells More Kindle Books than Hardcover and Paperback Combined." *Time Techland*. 19 May 2011. Web. 20 May 2011.

Glusac, Elaine. "American Airlines Tests In-Flight Streaming Video." *NYTimes.com*. 16 May 2011. Web. 21 May 2011.

Golijan, Rosa. "'Creepy' App Locates Anyone Using Photos." *Digital Life Today*. 22 June 2011. Web. 22 June 2011.

———. "What Happens on the Internet Every 60 Seconds." *Technolog*. *msnbc.msn.com*. 16 June 2011. Web. 22 June 2011.

Google Accounts Help. "Manage Your Online Reputation." *Google.com*. 16 June 2011. Web. 5 Oct. 2012.

Graser, Marc. "Amazon Bows 'Never before on DVD' Store." *Variety.com*. 24 May 2012. Web. 8 June 2012.

Gross, Doug. "Customers Fume over Netflix Changes." *CNN.com*. 20 Sept. 2011. Web. 21 Sept. 2011.

Gross, Melanie. "Netflix Accounts for More Web Traffic than Torrenting." *Ghacks.net*. 20 May 2011. Web. 21 May 2011.

Gruenwedel, Erik. "Disc Rental Profits Don't Lie at Netflix." *Home Media Magazine*. 30 Jan. 2012. Web. 1 June 2012.

———. "Streaming Helps CBS Up Entertainment Income 97%." *Home Media Magazine*. 2 Aug. 2011. Web. 3 Aug. 2011.

Hachman, Mark. "Fox Sues Dish over DVR Commercial Skipping Feature." *PCMag.com*. 24 May 2012. Web. 12 June 2012.

Hale, Mike. "A Parallel Universe to TV and Movies." *NYTimes.com*. 12 Nov. 2010. Web. 22 Feb. 2011.

Hampton, Tavis J. "The Future of Internet Video Streaming." *Pamilvisions.net*. 16 May 2011. Web. 21 May 2011.

Hardy, Quentin. "Active in Cloud, Amazon Reshapes Computing." *NYTimes.com*. 27 Aug. 2012. Web. 5 Sept. 2012.

Hayden, Erik. "Netflix Aims to Finish Off the DVD." *Atlantic Wire*. 23 Nov. 2010. Web. 3 Mar. 2011.

Hide, Nick. "Samsung Smart TV Watches You in Demo Video." *CNET UK*. 16 Jan. 2012. Web. 5 Mar. 2012.

Hope, Dan. "Netflix Streaming Makes up 22 Percent of Internet Traffic." *CSMonitor.com*. 18 May 2011. Web. 20 May 2011.

Horn, John. "The Reviewer Who Wasn't There." *Newsweek*. 2 June 2001. Web. 13 Aug. 2012.

Hosein, Hanson. "Storytelling, the New Studio System and the 'Death' of the Web." *Flipthemedia.com*. 3 Sept. 2010. Web. 13 June 2011.

Hurley, Mike. "We're About to Lose 1,000 Small Theaters that Can't Convert to Digital. Does It Matter?" *Indiewire.com*. 23 Feb. 2012. Web. 5 Mar. 2012.

International Business Times Staff. "Netflix Price Increase: 5 Reasons to Buy the Stock." *InternationalBusinessTimes.com*. 13 July 2011. Web. 4 Oct. 2012.

Ionescu, Daniel. "Google Co-Founder Spotted Wearing Google Glasses Prototype." *PCWorld.com*. 6 Apr. 2012. Web. 13 June 2012.

Ito, Joichi. "In an Open-Source Society, Innovating by the Seat of Our Pants." *New York Times*. 6 Dec. 2011, D9. Print.

Jackson, Eric. "Here's Why Google and Facebook Might Completely Disappear in the Next 5 Years." *Forbes.com.* 30 Apr. 2012. Web. 4 May 2012.

Jade, Kasper. "Apple Selling Half a Million Apple TVs per Quarter But No Update Planned for Q3." *AppleInsider.com.* 29 July 2011. Web. 4 Oct. 2012.

James, Meg. "Internet Distributors Are Changing the TV Syndication Game." *LATimes.com.* 31 Dec. 2011. Web. 5 Mar. 2012.

Kafka, Peter. "Netflix: We'll Do a Couple More *House of Cards*–Style Originals." *All Things D.com.* 25 Apr. 2011. Web. 25 Apr. 2011.

Kakutani, Michiko. "Love Found Amid Ruins of Empire." *NYTimes.com.* 26 July 2010. Web. 13 Aug. 2012.

Kastelein, Richard. "Netflix Streaming-Only Subscriptions in the Pipeline, But It Will Cost Them." *AppMarket.* 29 Oct. 2010. Web. 1 Mar. 2011.

Kehr, David. "Streaming Video's Emerging Bounty." *NYTimes.com.* 19 Aug. 2011. Web. 20 Aug. 2011.

Keller, Ed, and Brad Fay. "Facebook Can't Replace Face-to-Face Conversation." *USA Today.* 30 Apr. 2012, 9A. Print.

Kessler, Sarah. "Google Launches Tool for Online Reputation Management." *Mashable.com.* 16 June 2011. Web. 16 June 2011.

Kolakowski, Nicholas. "Apple Signing iTunes Cloud Music Deals: Reports." *Eweek.com.* 21 May 2011. Web. 21 May 2011.

Konow, David. "Netflix and Warner Bros. Keeping Obscure Films Alive." *TGdaily.com.* 19 May 2011. Web. 20 May 2011.

Kral, Georgia. "The Alley vs. the Valley: Why Qwiki Moved East." *MetroFocus.* 25 June 2012. Web. 26 July 2012.

Kravets, David. "Supreme Court Rules Congress Can Re-Copyright Public Domain Works." *Arstechnica.com.* 19 Jan. 2012. Web. 5 Mar. 2012.

Kung, Michelle, and Miguel Bustillo. "Wal-Mart to Give Hollywood a Hand." *WallStreetJournal.com.* 28 Feb. 2012. Web. 26 June 2012.

Lackey, Bill. "Rise of Redbox: Some Kiosks Are Located Outside Store Location in Parking Areas." *Selfserviceworld.com.* 20 Aug. 2007. Web. 4 Oct. 2012.

Lang, Brent. "Why America Doesn't Count at the Box Office Anymore." *Reuters.com.* 29 July 2011. Web. 31 July 2011.

Lasky, Julie. "Going beyond the Wand to the Brain." *New York Times.* 21 May 2012, D7. Print.

"Le Mans: Live! 24 Hours of Le Mans Video Streaming." *Speedtv.com.* 9 June 2011. Web. 10 June 2011.

Learmonth, Michael. "Soon Your TV Will Watch You, Too." *Adage.com.* 13 Jan. 2012. Web. 5 Mar. 2012.

Lee, Edmund, and Jonathan Erlichman. "Diller Says Aereo Will Expand to Most Large U.S. Cities." *Bloomberg.com.* 12 July 2012. Web. 26 July 2012.

Lieberman, David. "YOMYOMF Network Leaps into Top 10 in Deadline's YouTube Channels Ranking." *Deadline Hollywood.* 8 June 2012. Web. 4 Oct. 2012.

———. "Cable Execs Warn that Dish Network's Ad-Zapping DVR Could Lead to Higher Costs." *Deadline Hollywood.* 21 May 2012. Web. 12 June 2012.

"List of Massively Multiplayer Online Role-Playing Games." *Wikipedia.* 5 Oct. 2011. Web. 26 June 2012.

Listfield, Emily. "Generation Wired." *Parade.* 9 Oct. 2011, 9, 11, 14, 19. Print.

Lloyd, Robert. "Larry King Now Is Both New and Familiar." *LAtimes.com.* 17 July 2012. Web. 26 July 2012.

"Looxcie 2 Video Cam, 5-Hour." *SkyMall.* N.d., 104K. Print.

Madrigal, Alexis. "Sorry, Young Man, You're Not the Most Important Demographic in Tech." *The Atlantic.* 8 June 2012. Web. 4 Oct. 2012.

Maine, Charles Eric. *Escapement.* London: Hodder and Stoughton, 1956. Print.

Matheson, Whitney, and Alison Maxwell. "TV on the Web." *USA Today.* 17 June 2012, 6D. Print.

McClintock, Pamela. "23 Directors, Producers Speak out against Premium VOD." *Hollywoodreporter.com.* 20 Apr. 2011. Web. 20 Apr. 2011.

McCullagh, Declan. "How SOPA Would Affect You: FAQ." *CNET.* 21 Dec. 2011. Web. 26 June 2012.

McEwan, Ian. "10 Questions for Ian McEwan." *Time.com.* 7 June 2007. Web. 9 June 2012.

McGlaun, Shane. "Nielsen Looks at How People Are Streaming Hulu and Netflix." *Slash Gear.* 28 July 2011. Web. 4 Oct. 2012.

McKee, Steve. "The Profits and Perils of Viral Marketing," *Hotel Business Review.* 2008. Web. 4 Oct. 2012.

Morphy, Erika. "Heavily Traveled Roads to Netflix Could Face Traffic Jams." *E-CommerceTimes.com.* 19 May 2011. Web. 5 Oct. 2012.

Mui, Yian Q. "Redbox Finds Its Niche Focus on DVDs, Grocery Locations Fuel Growth." *Washington Post.* 28 Apr. 2007. Web. 4 Oct. 2012.

Mullaney, Tim. "Jobs Fight: Haves vs. the Have-nots." *USAToday.com.* 14 Sept. 2012. Web. 4 Oct. 2012.

Nakashima, Ryan. "Dish Sued over Ad-Skipping DVR Service." *USAToday.com.* 25 May 2012. Web. 12 June 2012.

"Negative Aspects of Social Networking." *Chennai Social Media.* 6 Oct. 2010. Web. 8 June 2012.

Newman, Jared. "Facebook Buys Face.com; Prepare for Easier Photo Tagging." *PCWorld.com.* 18 June 2012. Web. 18 June 2012.

Olsen, Mark. "Cinematography Nominees Discuss Film versus Digital." *LATimes.com.* 19 Feb. 2012. Web. 5 Mar. 2012.

Orlin, Jon. "How Google Pulled off Their Live Video Skydiving with Glasses DemoTechCrunch." *TechCrunch.com.* 28 June 2012. Web. 26 July 2012.

Ortutay, Barbara. "Google Creates a Spectacle." *Lincoln [NE] Journal Star.* 8 Apr. 2012, A8, A9. Print.

Oswald, Ed. "Security Firm Issues Alert on Facebook Facial Recognition." *PCWorld.com.* 7 June 2011. Web. 28 July 2011.

Pareles, Jon. "David Bowie, 21st-Century Entrepreneur." *NYTimes.com.* 9 June 2002. Web. 5 Mar. 2012.

Pedulla, Gabriele. *In Broad Daylight: Movies and Spectators after the Cinema.* Trans. Patricia Gaborik. New York: Verso, 2012. Print.

Pepitone, Julianne. "7% of Americans Subscribe to Netflix." *CNN.com.* 25 Apr. 2011. Web. 25 Apr. 2011.

Phend, Crystal. "Too Much TV Time Linked to Diabetes, Death." *ABCnews.com.* 15 June 2011. Web. 15 June 2011.

Potter, Ned. "Facebook Plans Server Farm in Sweden; Cold Is Great for Servers." *ABC News.* 27 Oct. 2011. Web. 4 Oct. 2012.

Prigg, Mark. "Facebook Share Price Hits All Time Low—Nearly Half Value at Float—After Revealing It Has More than 83 Million Fake Accounts." *DailyMail.co.uk.* 2 Aug. 2012. Web. 13 Aug. 2012.

Puente, Maria. "A-List Talents Get Caught up in Web." *USA Today.* 17 July 2012, D1. Print.

Purewal, Sarah Jacobsson. "Why Facebook's Facial Recognition Is Creepy." *PCWorld.com.* 8 June 2011. Web. 8 June 2011.

———. "Google Won't Dabble in Facial Recognition Search System." *PCWorld.com.* 19 May 2011. Web. 8 June 2011.

"Qwiki Arrives on Bing." *Bing.com.* 12 June 2012. Web. 26 July 2012.

Ramachandran, Shalini. "Zap! New DVR Wipes out Ads." *WSJ.com.* 11 May 2012. Web. 26 May 2012.

"Redbox Surpasses 100 Million DVD Rentals." *Kioskmarketplace.com.* 1 Feb. 2008. Web. 4 Oct. 2012.

Ressner, Jeffrey. "The Traditionalist." *DGA Quarterly.* Spring 2012. Web. 8 June 2012.

Richtel, Matt. "Silicon Valley Says Step away from the Device." *NYTimes.com.* 23 July 2012. Web. 26 July 2012.

———. "In Data Deluge, Multitaskers Go to Multiscreens." *NYTimes.com.* 7 Feb. 2012. Web. 5 Mar. 2012.

Roettgers, Janko. "Netflix Helps Boost Blu-ray Player and Disc Sales." *Gigaom.com.* 9 June 2011. Web. 9 June 2011.

Rose, Charlie. "George Lucas: Billionaire Down on Capitalism." *CBS News.* 20 Jan. 2012. Web. 8 June 2012.

Rutenberg, Jim. "The Gossip Machine, Churning out Cash." *New York Times.* 22 May 2011, 1, 20. Print.

Schiesel, Seth. "Expo Offers a Peek at Games of the Future." *New York Times.* 11 June 2012, C1, C2. Print.

Schonfeld, Erick. "What Apple Should Do with Its $100 Billion in Cash: Buy Hollywood." *TechCrunch.com.* 26 Jan. 2012. Web. 5 Mar. 2012.

Schumacher-Rasmussen, Eric. "Thousands of Channels and Everything Is On." *Streamingmedia.com.* June/July 2011. Web. 18 June 2011.

Seitz, Patrick. "Netflix Hits Record on Streaming Plan, by Mail Rate Hikes." *Investor's Business Daily.* 22 Nov. 2010. Web. 12 Mar. 2011.

Shefter, Milton, and Andy Maltz. *The Digital Dilemma: Strategic Issues in Archiving and Accessing Digital Motion Picture Materials.* Los Angeles: Academy of Motion Picture Arts and Sciences, 2007. Print.

Singer, Natasha. "Consumer Data, But Not for Consumers." *New York Times.* 22 July 2012, B3. Print.

———. "You for Sale: A Data Grant Is Mapping, and Sharing, the Consumer Genome." *New York Times.* 17 June 2012, B1, B8. Print.

"Siri—Frequently Asked Questions." *Apple.com.* 11 June 2012. Web. 5 Oct. 2012.

Sorrel, Charlie. "Word Lens: Augmented Reality App Translates Street Sign Instantly." *Wired.com*. 17 Dec. 2010. Web. 5 Mar. 2012.

Stelter, Brian. "Upending Anonymity, These Days the Web Unmasks Everyone." *NYTimes.com*. 20 June 2011. Web. 21 June 2011.

Stelter, Brian, and Bill Carter. "Networks Try a Social Spin at the Upfronts." *New York Times*. 19 May 2011, B7. Print.

Stelter, Brian, and Stuart Elliott. "At Network TV's Gathering, Web Is Central." *NYTimes.com*. 20 May 2012. Web. 22 May 2012.

Stone, Brad. "Amazon Erases Orwell Books from Kindle." *NYTimes.com*. 17 July 2009. Web. 26 June 2012.

Streitfeld, David. "Amazon Signs up Authors, Writing Publishers out of Deal." *NYTimes.com*. 16 Oct. 2011. Web. 8 June 2012.

Suddath, Claire. "What the World Will Look Like in 2050." *Time.com*. 30 June 2011. Web. 7 July 2011.

"Survey: Netflix Users Streaming More TV." *Nielsen.com*. 12 July 2012. Web. 26 July 2012.

Svensson, Peter. "Pay TV Industry Loses Record Number of Subscribers." *AP.org*. 10 Aug. 2011. Web. 10 Aug. 2011.

Tartaglione, Nancy. "UK Prime Minister David Cameron Calls for Focus on 'Commercially Successful' Pics." *Deadline.com*. 11 Jan. 2012. Web. 5 Mar. 2012.

Taylor, Jim. "UltraViolet FAQ." *UVdemystified.com*. 18 Feb. 2012. Web. 5 Oct. 2012.

Towner, Betsy. "Vanishing." *AARP Bulletin*. June 2012, 43. Print.

Tribbey, Chris. "Six Questions: Redbox's Mitch Lowe." *Home Media Magazine*. 31 July 2009. Web. 5 Oct. 2012.

Turkle, Sherry. "The Flight from Conversation." *New York Times*. 22 Apr. 2012, SR1, SR6. Print.

"Twisted Light Capable of Downloading 70 DVDs per Second." *Bunsen Burner*. 26 June 2012. Web. 26 June 2012.

Verrier, Richard. "4-D Movies: Next Big Thing for U.S. Theaters?" *LATimes.com*. 7 July 2012. Web. 26 July 2012.

———. "On Location: Birns & Sawyer Auctions Its Film Cameras." *LATimes.com*. 25 Oct. 2011. Web. 5 Mar. 2012.

Viane, Charles. Letter to Film Exhibitors. 30 June 2011. TS.

"Video Pen Is the Easiest, Stealthiest Way to Capture True Events." *SkyMall*. N.d., 29. Print.

Wauters, Robin. "Netflix: Our 20 Million Members Streamed 2 Billion+ Hours of Content in Q4 2011." *TechCrunch.com*. 4 Jan. 2012. Web. 8 June 2012.

Wohlsen, Marcus. "App Scans Faces of Bar-Goers to Guess Age, Gender." *AP News*. 19 May 2012. Web. 4 Oct. 2012.

Wortham, Jenna. "An Actor Who Knows Start-ups." *New York Times*. 26 May 2011, B1, B8. Print.

Wu, Tim. *The Master Switch: The Rise and Fall of Information Empires*. New York: Knopf, 2010. Print.

Yin, Sara. "Qwiki, Search Engine Funded by Facebook Co-Founder, Launches." *PCMag.com*. 24 Jan. 2012. Web. 26 July 2012.

"YouTube: Channel List." 26 June 2012. Web. 26 June 2012.

Index

Index

CPSIA information can be obtained at www.ICGtesting.com
Printed in the USA
LVOW07*1053030813

346080LV00003B/50/P